Use of Stable Isotopes of Carbon, Nitrogen, and Sulfur to Identify Sources of Nitrogen in Surface Waters in the Lower Susquehanna River Basin, Pennsylvania

By CHARLES A. CRAVOTTA III

Prepared in cooperation with the
PENNSYLVANIA DEPARTMENT OF ENVIRONMENTAL PROTECTION

U.S. GEOLOGICAL SURVEY WATER-SUPPLY PAPER 2497

[redacted]

U.S. DEPARTMENT OF THE INTERIOR
GALE A. NORTON, Secretary

U.S. GEOLOGICAL SURVEY
Charles G. Groat, Director

For sale by the
U.S. Geological Survey, Branch of Information Services
Box 25286
Denver, CO 80225-0286

Library of Congress Cataloging-in-Publication Data

Cravotta, Charles A.
Use of stable isotopes of carbon, nitrogen, and sulfur to identify sources of nitrogen in surface waters in the lower Susquehanna River Basin, Pennsylvania / by Charles A. Cravotta III.
p. cm. -- (U.S. Geological Survey water-supply paper : 2497)
"Prepared in cooperation with the Pennsylvania Department of Environmental Protection, Bureau of Soil and Water Conservation."
Includes bibliographic references (p. -).
Supt. of Docs. no.: I 19.13:W2497
1. Nitrogen--Environmental aspects--Susquehanna River Watershed. 2. Water--Pollution--Susquehanna River Watershed--Point source identification. 3. Radioactive tracers in water pollution research-Susquehanna River Watershed. I. Pennsylvania. Bureau of Soil and Water Conservation. II. Title. III. Series.
TD427.N5C73 1997 97-10381
628.1'61--dc21 CIP

ISBN 0-607-87207-1

CONTENTS

ILLUSTRATIONS

TABLES

CONVERSION FACTORS, VERTICAL DATUM, AND ABBREVIATIONS

Multiply	By	To obtain
	Length	
micrometer (µm)	0.00003937	inch
centimeter (cm)	0.3937	inch
meter (m)	3.281	foot
kilometer (km)	0.6214	mile
	Area	
hectare (ha)	2.471	acre
square kilometer (km^2)	0.3861	square mile
	Volume	
liter (L)	0.2642	gallon, U.S. liquid
	Flow rate	
liter per second (L/s)	15.85	gallon per minute
cubic meter per second (m^3/s)	35.311	cubic foot per second
	Mass	
kilogram (kg)	2.205	pound avoirdupois
megagram (Mg)	1.102	ton, short (2,000 lb)

Temperature in degrees Fahrenheit (°F) may be converted to degrees Celsius (°C) as follows:

$$°C = (°F - 32) / 1.8$$

Sea level: In this report, "sea level" refers to the National Geodetic Vertical Datum of 1929 (NGVD of 1929)—a geodetic datum derived from a general adjustment of the first-order level nets of both the United States and Canada, formerly called Sea Level Datum of 1929.

Abbreviated water-quality units used in this report:

- hour (hr)
- microsiemens per centimeter at 25 degrees Celsius (µS/cm)
- milligrams per kilogram (mg/kg)
- milligrams per liter (mg/L)
- parts per million (ppm)
- per mil (‰)
- volume percent (vol %)
- weight percent (wt %)
- specific conductance (S.C.)
- micrometer (µm)

Use of Stable Isotopes of Carbon, Nitrogen, and Sulfur to Identify Sources of Nitrogen in Surface Waters in the Lower Susquehanna River Basin, Pennsylvania

By Charles A. Cravotta III

ABSTRACT

Samples of nitrogen-source material, soil, and water were collected from several small, primarily single-source subbasins in the Lower Susquehanna River Basin during 1988–90 to determine the feasibility of using measurements of stable isotope ratios of carbon ($\delta^{13}C$), nitrogen ($\delta^{15}N$), and sulfur ($\delta^{34}S$) to identify sources of nitrogen (N) in stream water. Chemical and isotopic compositions were measured for six N-source types consisting of rain water, forest leaf litter, synthetic fertilizer, farm-animal manure, municipal-sewage effluent and sludge, and septic-tank effluent and sludge. Compositions of associated, nearby samples of topsoil, subsoil, runoff water, and stream water were measured to evaluate changes in compositions of transported N-containing materials near the N source. Animal manure, human waste (sewage plus septic), and forest-leaf litter can be distinguished on the basis of $\delta^{13}C$; however, most N-sources cannot be distinguished on the basis of $\delta^{15}N$ and $\delta^{34}S$, owing to wide ranges and overlap of compositions among different N-source types. Although values of $\delta^{15}N$ for soil and runoff-water samples are qualitatively similar to those of the applied N source, values of $\delta^{13}C$ and $\delta^{34}S$ for runoff-water and stream-water samples appear to reflect the compositions of relatively large reservoirs of the elements in soil organic matter and minerals, respectively, and not the composition of the applied N source. The ratio of organic carbon to total nitrogen (C-org:N), combined with $\delta^{13}C$, is useful for distinguishing agricultural soils, which have characteristically high $\delta^{13}C$ and low C-org:N, from forested soils. The C-org:N values of suspended particulates in runoff or stream waters generally are lower than those of nearby soils, however, and indicate that chemical transformations and resultant isotopic fractionation can be important controls on the compositions of N-containing compounds in the soil and water. In aqueous samples including surface water and liquid N-sources, isotopic ratios commonly differ between coexisting dissolved fractions of NO_3-N and NH_3-N and between dissolved and particulate fractions of N or S, probably because of isotopic fractionation during transport or N-source processing.

Isotopic measurements provide qualitative information about important reactions that can affect N concentrations in surface waters. However, mass-balance computations generally are not sufficiently accurate to estimate the proportions of multiple sources contributing to the N load in the streams studied because of (1) variations in source chemical and isotopic compositions and (2) nonconservative behavior and fractionation during transport over short distances (hundreds of meters). Uncertainties in mass-balance computations are complicated by the propagation of errors associated with measurements of discharge, chemical concentrations, and isotopic compositions of relatively dilute, small streams.

INTRODUCTION

The Susquehanna River is a major contributor of nutrients and sediment to Chesapeake Bay, the most productive estuary in North America. Excessive nitrogen (N) loading to the bay during the 20th century has caused eutrophication and anoxia (Ryther and Dunstan, 1971; Boynton and others, 1982; Officer and others, 1984). The N contamination results from human activities, primarily intensive agriculture and urban development, within the Lower Susquehanna River Basin (Ott and others, 1991) (fig. 1). Identifying the sources of N in downstream reaches of surface waters is complicated, however, because N compounds generally cannot be used as conservative tracers. Their transport and fate are affected by chemical transformations and uptake during transport and variable mobilities of different N species in gaseous, aqueous, and solid phases (Stevenson, 1972a, b; Brezonik, 1973; Hem, 1985). Consequently, a basic problem in controlling nutrient loads in the Susquehanna River Basin is the inability to distinguish among contributions from natural sources and various nonpoint and point N sources, including the atmosphere, fertilizers, animal wastes, and sewage.

Many natural and anthropogenic N sources also contain carbon (C) and sulfur (S). Each of the elements C, N, and S has at least two stable isotopes that exist in relatively constant proportions in the biosphere (table 1) (Fritz and Fontes, 1980). The isotopes of a particular element have slightly different mass-dependent properties that result in different rates of chemical reaction and partitioning among chemical species at equilibrium. The result of these differences is that the isotopes can be fractionated, or separated from one another by chemical and physical processes (Bigeleisen, 1965; Toran, 1982; Peterson and Fry, 1987; Coplen, 1993). Thus, the stable isotopic compositions of different compounds or species of a particular element can differ, and different sources of C, N, and S sometimes can be characterized on the basis of their isotopic compositions.

During 1988–90, the U.S. Geological Survey (USGS), in cooperation with the Pennsylvania Department of Environmental Protection (PaDEP), conducted a study to determine if a primary N source in selected subbasins of the Lower Susquehanna River Basin (fig. 1) could be identified by use of C, N, and S isotopic measurements of suspended-particulate and dissolved fractions in surface waters. Samples of N-source material, topsoil, subsoil, runoff water, and stream water were collected from different land-use areas in headwater reaches to evaluate changes in the composition of transported N-bearing materials near the N source. Six locally important N sources were considered including rainwater, forest-leaf litter, synthetic fertilizer, farm-animal manure, human septic waste, and urban sewage.

Purpose and Scope

This report presents the results of chemical and isotopic analyses of the N-source materials and nearby soil and water samples and discusses the use of stable C, N, and S isotopic tracers to determine N contributions from different sources. Three basic hypotheses are tested: (1) C, N, and S isotopic compositions of different N-source materials differ, (2) isotopic compositions of suspended-particulate and dissolved fractions in aqueous N-source and surface-water samples differ, and (3) isotopic compositions of applied N sources and nearby soil, runoff-water, and stream-water samples are similar. Additionally, results of isotopic mass-balance and fractionation computations are used to estimate N loads contributed by multiple sources and to explain isotopic variability in selected subbasins resulting from N-transformation processes.

Acknowledgments

The author is indebted to Brian Fry, Robert "Happ" Garritt, Kris Tholke, Wendy Cochran, and Robert Michener (presently at Boston University) of the Marine Biological Laboratory, Ecosystems Center, Woods Hole, Mass., for performing isotopic analyses, and to Timothy Bergstresser, Dave Statler, and coworkers of the Geochemical Testing Laboratory, Somerset, Pa., for performing chemical analyses. The author is grateful to Carol Kendall of the USGS for providing helpful advice and access to her reference files. Special thanks are due to the following landowners and property managers who permitted access to the sampling locations and provided information: Kenny Moore (Bald Eagle), Stanley Herr (Brush Run), Titus Zimmerman (Conestoga FS#1), Clark Stauffer (Conestoga FS#2), John S. Smith (York Wastewater Treatment Plant), Fran McGovern (Harrisburg

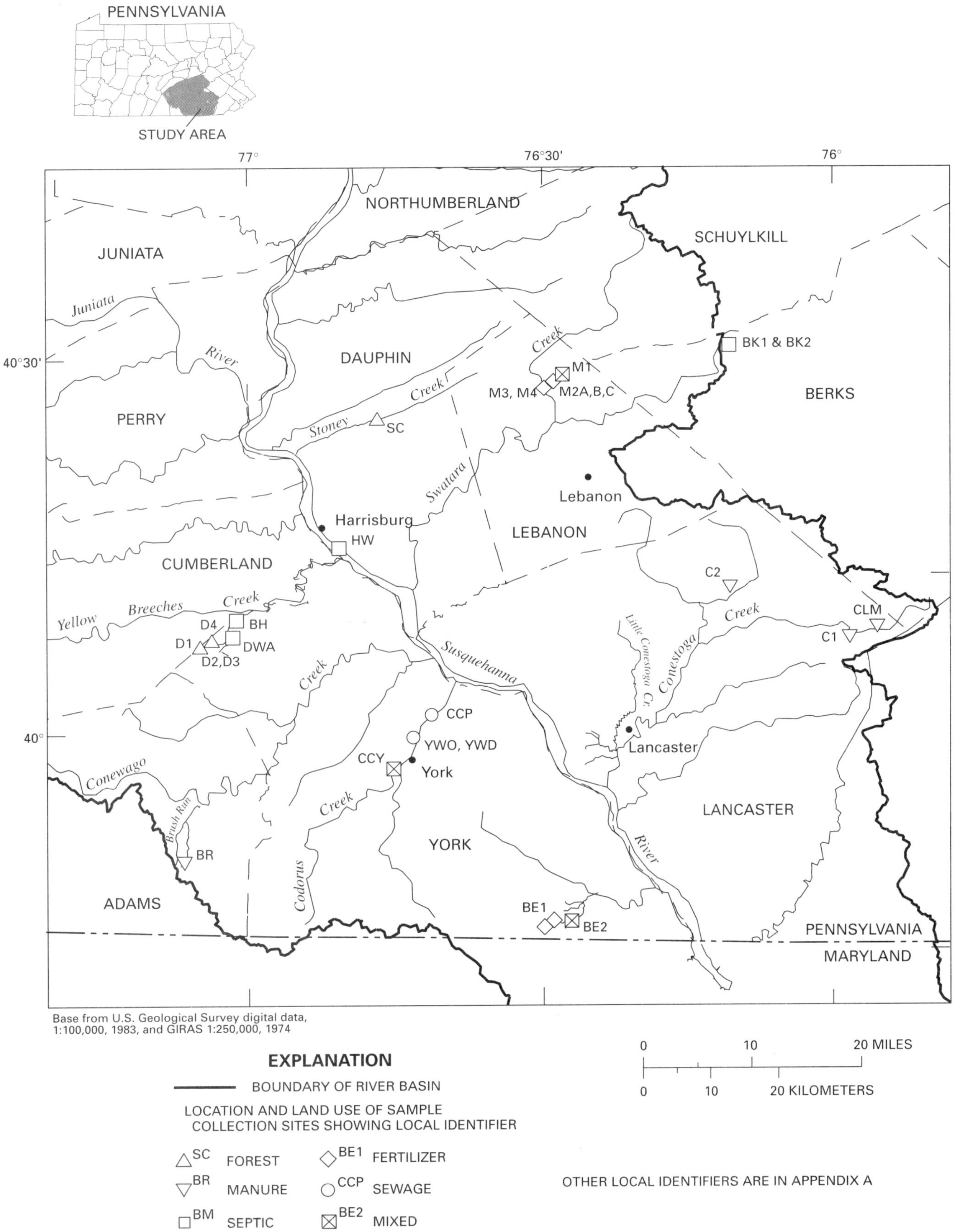

Figure 1. Location of sample-collection sites in the Lower Susquehanna River Basin. (Detailed site descriptions are given in Appendix A.)

Table 1. Geochemical characteristics of carbon, nitrogen, and sulfur[1]

[ppm, parts per million; amu, atomic mass units; δ, isotopic ratio delta value; ‰, per mil]

	Carbon (C)	Nitrogen (N)	Sulfur (S)
Atomic number (Atomic weight, amu):	6 (12.011)	7 (14.0067)	16 (32.06)
Stable isotopes (Abundance, percent)	^{12}C (98.89) ^{13}C (1.11)	^{14}N (99.634) ^{15}N (.366)	^{32}S (95.02) ^{34}S (4.21)
Common chemical forms			
Gaseous compounds	CO_2, CO, CH_4	NO_2,N_2O, N_2, NH_3	SO_2, H_2S
Aqueous species	$H_2CO_3^0$, HCO_3^-, CO_3^{2-}, CH_4	N_2, NO_3^-, NO_2^-, NH_4^+	SO_4^{2-}, H_2S^0, HS^-
Mineral compounds	$CaCO_3$, $CaMg(CO_3)_2$	KNO_3, NH_4^-EX	$CaSO_4 \cdot 2H_2O$, FeS_2
Organic compounds	$C_6H_{12}O_6$ = carbohydrates	$CO(NH_2)_2$ = urea	$HSCH_2CH(NH_2)COOH$ = cystein
Typical abundance (ppm):			
Atmosphere	CO_2 322 to 332	N_2 780,900	SO_2 0.0002
Freshwater	HCO_3^- C 2 to 30	NO_3^- N .1 to 5	SO_4^- S 1 to 30
Soils	4,000 to 120,000	440 to 5,440	100 to 2,000
Plants	450,000 to 500,000	2,000 to 55,000	100 to 800
Isotopic composition (δ, ‰):			
Atmosphere	CO_2 −6 to −8	N_2 0	+1 to +7
Freshwater	HCO_3^- C −15, POM −35	NO_3^- N +4 to +7	−22 to +20
Soils	−18 to −31	−4 to +14	−15 to +25
Plants	−12 to −30	−8 to +2	−10 to +22
Isotopic standard reference:	Pee Dee Formation belemnite ($CaCO_3$)	Atmospheric gas nitrogen (N_2)	Canyon Diablo troilite (FeS)
Standard abundance ratio:	$^{12}C/^{13}C$ = 88.99	$^{14}N/^{15}N$ = 272.0	$^{32}S/^{34}S$ = 22.22

[1]Sources: Coplen (1993), Coplen and others (1992), Peterson and Fry (1987), Hem (1985), Olson and Kurtz (1982), Toran (1982), Berg and Staaf (1981), Fritz and Fontes (1980), Faure (1977), Brezonik (1973), Kaplan (1972), Pearson and Rightmire (1980), Field (1972), Stevenson (1972a, b; 1982), Thode (1972), Thode and others (1961).

Wastewater Treatment Plant), Chad Wagner (Dillsburg Borough Water and Sewer Department), Gary Morrow (Monroe Township Sewage Enforcement Officer), and Randy Olinger (Monroe Valley Golf Course).

TERMINOLOGY AND PRINCIPLES OF ISOTOPIC MEASUREMENTS

The environmentally significant stable isotopes of C, N, and S, common chemical forms, and abundance of C, N, and S in the atmosphere, freshwater, soils, and plants are summarized in table 1. Important biochemical transformations of C, N, and S are shown in figure 2. General reviews of the terminology, measurement, and natural variations of C, N, and S isotopes are presented by Fritz and Fontes (1980), Toran (1982), Peterson and Fry (1987), and Coplen (1993).

Isotope-ratio analysis involves precise measurement, usually by mass spectrometry (Bowen, 1988), of the more abundant light isotope relative to the less abundant heavy isotope (for example, $^{13}C/^{12}C$, $^{15}N/^{14}N$, and $^{34}S/^{32}S$) in carbon dioxide (CO_2), nitrogen gas (N_2), or sulfur dioxide (SO_2) gas generated from combustion of the sample material. This ratio is reported relative to the isotopic ratio in a reference standard (table 1) (Fritz and Fontes, 1980; Mariotti, 1983; Peterson and Fry, 1987; Bowen, 1988). The isotopic composition is expressed in terms of the isotopic ratio delta value (δ), in per mil, defined as

$$\delta E = [(R_{sample}/R_{standard}) - 1] \cdot 1{,}000 \qquad (1)$$

where E is an element (in this report C, N, or S) and R is the ratio of $^{13}C/^{12}C$, $^{15}N/^{14}N$, or $^{34}S/^{32}S$ in the sample or standard.

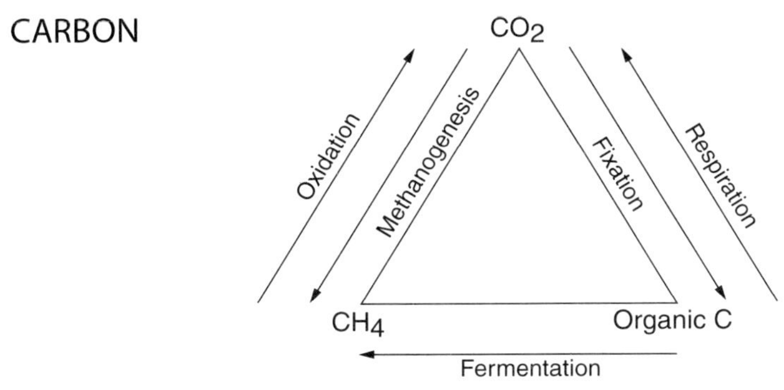

C-cycle processes:

Respiration:	C-org	$\rightarrow CO_2$
Oxidation:	CH_4	$\rightarrow CO_2$
Fermentation:	C-org	$\rightarrow CH_4$
Methanogenesis:	CO_2	$\rightarrow CH_4$
*Fixation:	CO_2	$\rightarrow$C-org

*(Photosynthesis or chemosynthesis)

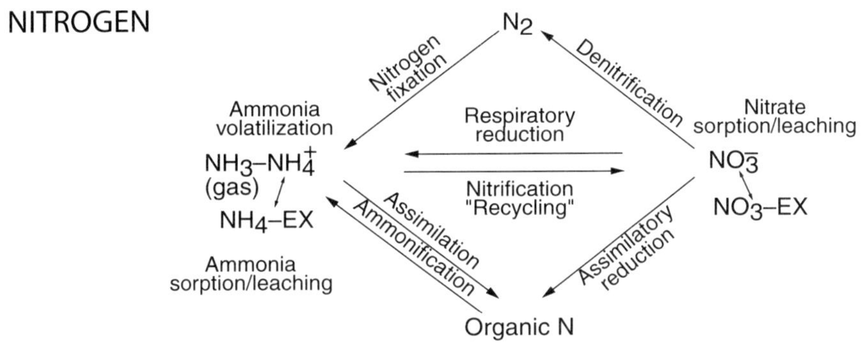

N-cycle processes:

Nitrogen fixation:
$N_2 \rightarrow NH_4^+ \rightarrow$N-org

Ammonification:
N-org $\rightarrow NH_4^+$

Nitrification:
$NH_4^+ \rightarrow NO_2^- \rightarrow NO_3^-$

Assimilation:
$NH_4^+ \rightarrow$N-org

Respiratory reduction:
$NO_3^- \rightarrow NO_2^- \rightarrow NH_4^+$

Denitrification:
$NO_3^- \rightarrow NO_2^- \rightarrow NO \rightarrow N_2O \rightarrow N_2$

Volatilization:
$NH_4^+ \rightarrow NH_3$ gas

Ammonium sorption/leaching:
NH_4^+ aq $\leftrightarrow NH_4$-EX

Nitrate sorption/leaching:
NO_3^-aq $\leftrightarrow NO_3$-EX

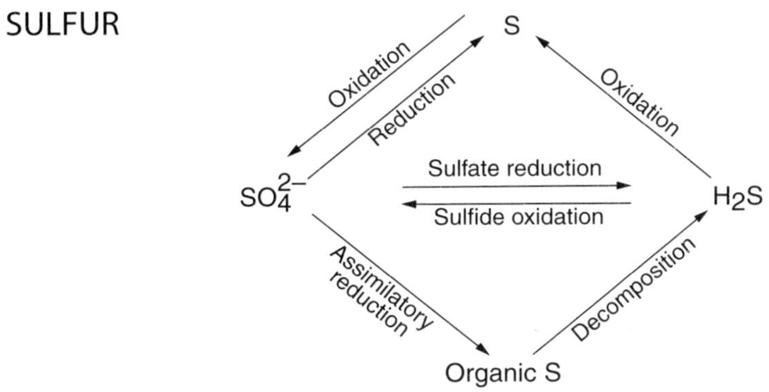

S-cycle processes:

Oxidation:
$H_2S \rightarrow S^0 \rightarrow SO_4^{2-}$
$H_2S \rightarrow SO_4^{2-}$

Sulfate reduction:
$SO_4^{2-} \rightarrow H_2S$

Assimilatory reduction:
$SO_4^{2-} \rightarrow$S-org

Decomposition:
S-org $\rightarrow H_2S$

Figure 2. Biochemical cycles of carbon, nitrogen, and sulfur.

Isotopic Mass Balance

Isotopic compositions expressed as delta values are additive, such that the isotopic composition of the reactant must equal that of the products when summed in stoichiometric proportions. If δ is the isotopic composition and Q the mass, then the mass and isotopic balances are, respectively,

$$Q_r = Q_a + Q_b \tag{2}$$

and

$$\delta_r \bullet Q_r = (\delta_a \bullet Q_a) + (\delta_b \bullet Q_b) \tag{3}$$

Equations 2 and 3 apply to stoichiometric chemical reactions, for example, where Q_r is the moles of N in a reactant, and Q_a and Q_b are the moles of N in the products. The equations also apply to simple mixing of two N-containing materials or waters (Q_a and Q_b, where $Q = V \cdot c$, and V is volume and c is concentration) having different isotopic compositions (δ_a and δ_b) to produce the final mixture (Q_r) (Krouse, 1980; Mariotti and others, 1981, 1988); the mixture will have an intermediate isotopic composition (δ_r) depending on the relative contributions of added materials. Equations 2 and 3 can be combined as

$$Q_b = Q_r \bullet \left(\frac{\delta_r - \delta_a}{\delta_b - \delta_a}\right) \tag{4}$$

or

$$\delta_b = \frac{(\delta_r \bullet Q_r) - (\delta_a \bullet Q_a)}{Q_r - Q_a} \; . \tag{5}$$

Equation 4 can be used to estimate the N load from a nonpoint N source (Q_b) contributing to the measured total N load at a downstream point (Q_r), if isotopic compositions of upstream (δ_a), N-source (δ_b), and downstream (δ_r) samples are known. Equation 5 can be used to estimate the isotopic composition of the N source (δ_b), if the loads (Q_a, Q_r) and isotopic compositions (δ_a, δ_r) at upstream and downstream points, respectively, are known.

Isotopic Fractionation

Fractionation during equilibrium (reversible) or disequilibrium (unidirectional) processes results because atomic masses and bond strengths differ for different isotopes. Isotopic equilibrium exchange reactions involve redistribution of isotopes of an element among phases or chemical species (Coplen, 1993). At isotopic equilibrium, the forward and backward reaction rates of the lighter isotopic species or molecules are equal and those of the heavier isotopic species or molecules are equal. For example, during equilibrium, volatilization, or dissolution of gases such as CO_2 and ammonia (NH_3), the heavier isotope tends to concentrate in the aqueous phase because the lighter isotope has a higher vapor pressure (Bigeleisen, 1965). Although the isotopic ratios in the aqueous and gaseous phases differ at equilibrium, the ratios in the two phases vary in constant proportion. Equilibrium processes generally take place in closed or semiclosed systems.

Kinetic fractionation can result in nonequilibrium systems in which reaction rates are mass dependent. As a general rule, the lighter isotope reacts faster than the heavier isotope (Coplen, 1993). For example, during evaporation or sublimation, the system is open, and the volatile, isotopically lighter product can escape, which leads to wide variations in delta values of the product and residual reactant. Most biologically mediated reactions are unidirectional, resulting in isotopically heavier reactants and isotopically lighter products during the course of a reaction (Letolle, 1980; Peterson and Fry, 1987; Coplen, 1993).

During a single-step, unidirectional reaction, the isotopic composition of the reactant and instantaneously formed product is a simple function of the progress of the reaction in accordance with the following Rayleigh equation (Krouse, 1980; Mariotti and others, 1981, 1988; Peterson and Fry, 1987):

$$\delta_r = \delta_{ro} - D_{r/p} \bullet ln\left(\frac{c_r}{c_{ro}}\right) \tag{6}$$

where c_{ro} and c_r are the reactant concentration at time t = 0 and time t, respectively, and δ_{ro} and δ_r are the isotopic composition of the reactant at time t = 0 and time t, respectively. $D_{r/p}$ is the isotopic discrimination of the reaction, which is related to the isotopic kinetic fractionation factor, $\alpha_{r/p}$ (= $^{13}C/^{12}C$, $^{15}N/^{14}N$, or $^{34}S/^{32}S$ in the residual reactant divided by that in the product):

$$D_{r/p} = 1,\ 000 \bullet (\alpha_{r/p} - 1) \, . \tag{7}$$

Values of $D_{r/p}$ are positive in sign when the lighter isotope reacts faster than the heavier isotope and can be closely approximated as the per mil difference between an instantaneous product and reactant (Hubner, 1986; Peterson and Fry, 1987; Coplen, 1993):

$$D_{r/p} \cong \delta_r - \delta_p \,. \qquad (8)$$

Such approximate values of $D_{r/p}$ have been determined by previous investigators to derive apparent kinetic fractionation factors ($\alpha_{r/p}$) for many of the N transformation reactions in soils and waters (fig. 2) (Kreitler, 1975; Letolle, 1980; Hubner, 1986). Thus, if the extent of the transformation reaction and the corresponding fractionation factor are known, isotopic effects from fractionation may be computed by use of a combined form of equations 5 and 6

$$\delta_r = \delta_{ro} - 1,\ 000 \bullet (\alpha_{r/p} - 1) \bullet ln\left(\frac{c_r}{c_{ro}}\right) \,. \qquad (9)$$

Figure 3 was constructed on the basis of equations 6 and 9 to show the effect of processes having fractionation factors ($\alpha_{r/p}$) greater than 1.0, which is appropriate for most N-cycle processes. When a small amount of reactant has been converted to the product, both the accumulated and instantaneous products are depleted in the heavier isotope and have similar delta values. As the reaction proceeds, (1) the remaining reactant, instantaneous product, and accumulated product become progressively more enriched in the heavier isotope, and (2) the per mil difference becomes larger between the remaining reactant and the accumulated product and smaller between the remaining reactant and the instantaneous product. When all the reactant is consumed, the accumulated product has the isotopic composition of the initial reactant ($\bar{\delta}_p = \delta_{ro}$).

USE OF STABLE ISOTOPES AS TRACERS

The use of C, N, and S isotopes to identify N sources is based on the concept that these elements are interrelated in the biochemical N cycle (Stumm and Morgan, 1981; Bolin and Cook, 1983; Peterson and Fry, 1987), and that measurable differences in the isotopic composition of N-source materials will persist as N-containing compounds are transported from the source. The isotopic compositions and forms of C, N, and S in soil and water may resemble those of a nearby N source. However, the composition of soils and waters not only reflects the composition of the original source, or of mixed sources having different compositions (for example, biologically fixed N in soil, synthetic fertilizer, and animal waste), but can be influenced by isotopic fractionation during the transport and chemical transformation of C, N, and S compounds. Thus, the $\delta^{13}C$, $\delta^{15}N$, and $\delta^{34}S$ values of the material from which a compound formed establish an isotopic "baseline" that can be subsequently shifted by isotopic fractionation.

For isotopes to be most useful as tracers of N sources, fractionation should occur prior to transport, causing sources to have unique isotopic ratios, and fractionation should be minimal during transport from the source to nearby surface waters, so the transported products will inherit the source isotopic ratios. The partial loss of volatile species formed under reducing conditions [methane (CH_4), NH_3, N_2, hydrogen sulfide (H_2S)] can cause major fractionations in C, N, and S in most anthropogenic N sources (Toran, 1982). In contrast, aerated, free-flowing streams may be ideal for retaining original isotopic values during transport of particulate matter and oxidized solutes. Particulates generally are nonreactive compared to most dissolved species. Ideally, suspended particulates in stream water should consist of small fragments of the original N-source material(s) and should have an isotopic composition similar to the source. Interpretation of isotopic compositions of particulates can be complicated, however, because biological processes can add particulates to the suspended load. Algae, plankton, and bacteria can compose a substantial part of the particulate load (Berg and Staaf, 1981; Fairchild and others, 1983; James and others, 1988), especially during low-flow conditions in the summer and fall.

Carbon

The most important C forms in the biosphere are gaseous CO_2 and CH_4, dissolved CO_2 (carbonate species), solid carbonate minerals, and organic compounds (table 1). Major biochemical C-cycle processes (fig. 2) include photosynthesis and chemosynthesis, whereby CO_2 is converted into organic matter; respiration, whereby organic compounds are oxidized to CO_2; and methanogenesis or fermentation, which may be considered reduction of

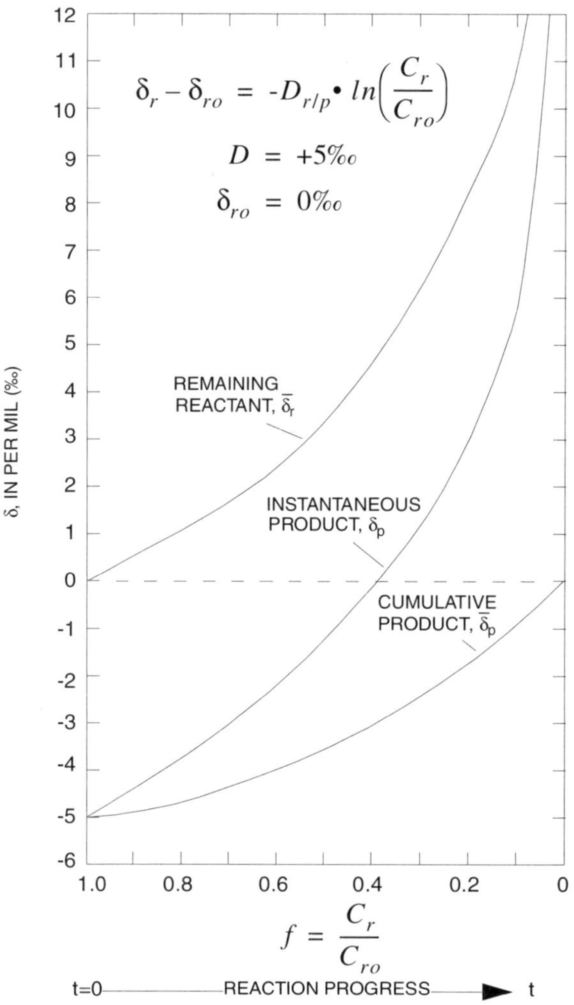

Figure 3. Effect of kinetic fractionation on isotopic compositions of reactant and product. [Curves are based on Rayleigh distillation, equation 6 in text, and show theoretical evolution of isotopic compositions of components during a single-step, first-order process where the lighter isotope reacts faster than the heavier isotope (α >1) and f ($=\frac{C_r}{C_{ro}}$) is the proportion of reactant remaining. The upper curve indicates the composition of the remaining reactant ($\bar{\delta}_r$), which is well mixed. The middle curve indicates the composition of the instantaneous product formed (δ_p). The lower curve indicates the composition of the accumulated product ($\bar{\delta}_p$), if well mixed.]

CO_2 to CH_4 (Bolin, 1970; Stumm and Morgan, 1981). The most important factor affecting C-isotopic compositions of natural compounds in the biosphere is the effect of absorption and photosynthetic fixation of CO_2 by plants (Bender, 1972; Deines, 1980; Nadelhoffer and Fry, 1988; Sackett, 1989). Photosynthesis by upland trees and northern grasses involves a net fractionation of about 19‰, whereas that by tropical grasses including corn and maize involves a small fractionation of about 6‰ (Park and Epstein, 1960; Smith and Epstein, 1971; Deines, 1980). Additional biological mechanisms for fractionation of C isotopes include microbial decay processes, such as the formation of CH_4 during anaerobic decomposition (Baedecker and Back, 1979; Toran, 1982) and of CO_2 during aerobic respiration (Balesdent and others,

1988; Nadelhoffer and Fry, 1988). These processes enrich the product gases in ^{12}C and can leave the organic-C reactant enriched in ^{13}C.

Values of $\delta^{13}C$ are expected to be smaller for forest vegetation than for corn and maize owing to differences in fractionation during photosynthesis. Farm animals such as cattle, swine, and poultry eat a diet consisting of corn grain and corn fodder plus smaller amounts of other materials. Diet is the primary determinant of animal C-isotopic compositions (Peterson and Fry, 1987), and herbivore excrement is similar in $\delta^{13}C$ relative to that of the plant diet (Teeri and Schoeller, 1979). Thus, $\delta^{13}C$ values of natural leaf litter and farm-animal manure are likely to differ because farm-animal manure will reflect a corn diet. Accordingly, the $\delta^{13}C$ values for topsoil in a forested watershed are likely to differ from those for soils in an agricultural watershed where corn is the principal crop or where animal manure is the principal N source. The C-isotopic composition of suspended particulates in waters draining forested or agricultural subbasins may be relatively unchanged chemically and isotopically from the original source material.

Nitrogen

The most important N forms in the biosphere are N_2; dissolved nitrate (NO_3^-), nitrite (NO_2^-), ammonium (NH_4^+), and organic-N; mineral-fixed NH_4^+; and organic-N compounds (Delwiche, 1970; Stevenson, 1972b; Sprent, 1987). Naturally occurring organic N consists primarily of amino and amide (proteinaceous) N along with some heterocyclic compounds present as cellular constituents, as nonliving particulate matter, and as soluble organic compounds (Brezonik, 1973). Major biochemical N-cycle processes (fig. 2) include N_2-fixation, ammonification, nitrification, uptake or assimilation, respiratory nitrate reduction, and denitrification. With the exception of ammonium assimilation, each biochemical transformation involves a change in redox state of N; all involve a change in pH (Sprent, 1987). Major physicochemical processes include ammonia volatilization and sorption. With the exception of adsorption reactions, the N-cycle processes tend to cause depletion of the heavier isotope in the products relative to the reactants (Kreitler, 1975; Letolle, 1980; Hubner, 1986).

The $\delta^{15}N$ values for natural soil N from biological N_2 fixation and synthetic fertilizers are similar to that of atmospheric N_2 (Shearer and others, 1974, 1978; Freyer and Aly, 1974; Hubner, 1986); however, $\delta^{15}N$ values for residual N and NO_3^- derived from animal and human excrement are larger owing to volatilization of NH_3 (Kreitler, 1975; Gillham and others, 1978; Wolterink and others, 1979; Letolle, 1980). Thus, in attempts to identify sources of N contamination in water supplies, many investigators have measured $\delta^{15}N$ in different N sources and associated samples of soil, surface water, and ground water at various localities (Kreitler and Jones, 1975; Kreitler, 1975, 1979; Kreitler and others, 1978; Kreitler and Browning, 1983; Gormly and Spalding, 1979; Wolterink and others, 1979; Mariotti and others, 1980, 1984, 1988; Spalding and others, 1982; Flipse and others, 1984; Flipse and Bonner, 1985; Heaton and others, 1983; Heaton, 1984, 1986; Exner and Spalding, 1985; Komor and Anderson, 1993; Aravena and others, 1993). In general, previous work has consistently indicated that soil and ground-water NO_3-N derived from animal and human wastes cannot be distinguished from one another on the basis of their $\delta^{15}N$, but they can be distinguished from NO_3-N derived from natural soil N, fertilizer, and munitions. Consequently, N isotopic analysis can aid in distinguishing among NO_3-N from animal (including human) wastes, synthetic fertilizer, and natural N sources.

Sulfur

The most common S forms in the biosphere include gaseous oxides (SO_x); dissolved sulfate (SO_4^{2-}), sulfides (H_2S, and HS^-); mineral sulfates and sulfides; and organic-S compounds (table 1). S and N are present in proteins, which are composed of S-containing amino acids such as cysteine, cystine, and methionine (Field, 1972). Major biochemical S-cycle processes (fig. 2) include microbial oxidation of organic S, native S, and sulfides to SO_4^{2-}; assimilatory reduction of SO_4^{2-}; and dissimilatory reduction of SO_4^{2-} (Thode, 1972; Krouse, 1980). The most important processes affecting S isotopic compositions of natural compounds are kinetic fractionation accompanying the reduction or oxidation of S compounds (Thode and others, 1961; Thode, 1972; Kaplan and Rittenberg, 1964; Kaplan, 1972;

Chambers and Trudinger, 1979; Krouse, 1980; Pearson and Rightmire, 1980; Stam and others, 1992). These processes tend to cause depletion of the heavier isotope in the products relative to the reactants. Abiotic reduction or bacterial dissimilatory reduction of sulfate or sulfite produces sulfide depleted in ^{34}S. Fractionation during sulfide oxidation can produce sulfate depleted in ^{34}S. However, the fractionation effect from reduction processes generally is greater than that from oxidation processes (Krouse, 1980; Pearson and Rightmire, 1980; Peterson and Fry, 1987).

The $\delta^{34}S$ of SO_4^{2-} from evaporite minerals generally differs from that of SO_4^{2-} from oxidized organic compounds and sulfide minerals because of effects from SO_4^{2-} reduction and because of differences in source compositions. Hence, the $\delta^{34}S$ of septic-tank effluent and sewage-treatment effluent may differ because septic effluent is anaerobic and produces sulfide that can be precipitated in sludge or volatilized. In contrast, sewage effluent is aerated and produces SO_4^{2-}, which is soluble.

DATA COLLECTION AND ANALYSIS

Study-Site Selection

Study sites were selected in 11 subbasins of the Lower Susquehanna River Basin that were in headwater areas and that each contained one dominant type of land use (fig. 1). Sites that were already instrumented and monitored for other USGS studies were given preference. In order to increase the transferability of data gathered, at least two sites having similar land use or similar N sources were sought. Thus, two forested watershed sites (Stony Creek and upper Dogwood Run), two fertilizer-use sites (Monroe Creek and Bald Eagle Creek), three manure-use sites (Brush Run and Conestoga River field sites), two septic-field-use sites (middle Dogwood Run and Berkshire Hills), and two sewage-treatment-discharge sites (lower Dogwood Run and Codorus Creek) were selected (fig. 1, table A1) for sampling of N-source materials, soils, and waters. In addition, the Dillsburg, Harrisburg, and York sewage-treatment (wastewater) plants were selected for sampling of wastewater and sludge from mixed domestic and industrial sources.

Sampling locations within each land-use area were determined after study of topographic maps that indicated directions of surface-water flow. Soil-sampling locations were from areas where the N-source material was applied at the surface 3 months or less prior to sampling. Runoff-water and stream-water sampling locations were no more than 100 m downgradient from soil-sampling locations or source-sampling locations, with only one exception (Codorus Creek at Pleasureville). Two general categories of stream-water sampling locations were established. At "single-source" sites, one predominant N source could be identified; at "mixed-source" sites, more than one N source could be identified. Many of the mixed-source sites were located downstream from single-source sites. Appendix A provides detailed descriptions of the subbasins and sampling sites.

Sample Collection and Processing

N-source materials, soil, and surface-water samples were collected from single-source or mixed-source sites over relatively short time intervals (1 day per site; 2 weeks total elapsed) during low- and high-flow periods in 1988–90 (table B1). N-containing materials thought to be locally important sources were collected, including atmospheric precipitation from several storm events; animal manure from cattle, swine, and chicken; synthetic fertilizers; and human sewage from septic systems and municipal wastewater-treatment plants (table B1).

Precipitation water was collected into clean 4-L polyethylene jugs fitted with plastic funnels that were set up in ice-filled coolers placed in field locations at the start of selected rainstorms. The water was retrieved the next day, after the storm had passed. The cumulative rainfall quantity that fell during the 24-hour period, and measurements of specific conductance (S.C.), and pH of the water were recorded (table B2), and then the samples were acidified with reagent-grade hydrochloric acid (HCl) to 0.5 vol % HCl concentration. Because the precipitation water was dilute (S.C. less than 35 μS/cm), 1-L volumes were boiled down to 0.25-L volumes in an effort to concentrate solutes for N isotopic analysis.

Manure was collected as grab samples from storage "lagoons," animal pens, and the soil surface where it had been spread on fields. Fertilizer also was collected as grab samples from materials stored in the vicinity of sample sites. Wastewater-treatment sludge and effluent were collected as grab samples at the treatment plants. Septic-system samples were collected from two different pump trucks that were delivering septic waste to the Harrisburg treatment plant and also from an access pipe in a septic field. In order to prevent microbial activity and NH_3 volatilization, the N-source samples were acidified with HCl in the field. Aqueous samples of swine manure, septic effluent, and sewage effluent were acidified to 0.5 vol % HCl concentration after pH and S.C. were measured in the field (table B2) and were stored on ice until they could be processed further as described below for water samples. The solid manure, fertilizer, and sludge samples were oven-dried at 60° C for a minimum of 8 hours. The dry samples were pulverized with a mortar and pestle to a 250-μm mesh size, and then split into subsamples by use of a Soiltest sample splitter. The subsamples were wrapped in aluminum foil and frozen until analyzed. (Because the aluminum foil eventually decomposed owing to the acid in samples that were analyzed, it became necessary to transfer archived subsamples to glass vials.)

Samples of surface soil (0–8 cm) and subsurface soil (8–16 cm), hereinafter called topsoil and subsoil, respectively, were collected twice at each forested, manure-use, or fertilizer-use site. Soil samples were collected by use of a stainless-steel spoon and auger at 12 locations that were spaced at roughly 10-m intervals along 2 arbitrary, perpendicular traverses across each site. The samples were composited in the field and stored in baked glass jars. The composite samples were acidified with HCl and then processed as described above for the solid manure and fertilizer samples.

Stream-, spring-, and runoff-water grab samples were collected into baked wide-mouth glass bottles or 4-L polyethylene jugs suspended below the water surface. A total volume of at least 12 L collected at each location was composited in a churn, transferred to 4-L polyethylene jugs, and acidified to 0.5 vol % HCl concentration. Stream discharge, temperature, pH, and S.C. were measured at the time of sampling (table B2). Water samples were then stored on ice until they were processed in the laboratory.

The acidified aqueous samples were divided in the laboratory into "whole," "dissolved," and "particulate" fractions for analysis. To obtain "dissolved" and "particulate" fractions, sample volumes of 1–12 L were vacuum-filtered through baked, 1.5-μm-pore-size, Whatman 934-AH glass-fiber filters until at least three filters clogged with particulate material. A 3-L portion of the filtrate and a 3-L portion of the original whole sample were transferred to 1-L polyethylene bottles and then frozen until analyzed; the particulate-clogged filters were oven-dried at 60° C for a minimum of 8 hours and then wrapped in aluminum foil and frozen until analyzed.

Laboratory Analysis

Frozen subsamples of the solid and aqueous samples were shipped on ice to two different, private laboratories for analysis of C, N, and S chemical concentrations and isotopic ratios, respectively.

Concentrations of the analytes were measured by the methods indicated in table 2. Concentrations of total C, organic C, and inorganic C were measured in solid samples and unfiltered aqueous samples but not in the filtered aqueous samples. Concentrations of organic-N, NH_3-N, and NO_3-N were measured in solid and filtered and unfiltered (whole) aqueous samples; concentrations of NO_2-N also were measured in aqueous samples but not in solids. Concentrations of organic S, SO_4-S, and sulfide-S were measured in solids, but only SO_4-S and sulfide-S were measured in aqueous samples.

Stable C, N, and S isotope ratios in samples were measured as the gases CO_2, N_2, and SO_2 on Finnigan MAT 251 or Delta S isotope-ratio mass spectrometers. Solid and particulate samples for C and N isotopic ratios were analyzed by use of an automated system for combustion, reduction, and cryogenic trapping of pure CO_2 and N_2 as described by Fry and others (1992). Although this method allows for the determination of $\delta^{15}N$ for the total N in each sample, particulates analyzed were assumed to contain predominantly organic N. Aqueous samples for N isotope ratios of NO_3^- and NH_4^+ were steam distilled and analyzed as sorbed NH_4^+ on a zeolite molecular sieve according to methods of Velinsky and others (1989) and Horrigan and others (1990). Dissolved NO_3^- was converted to NH_4^+ with Devarda's alloy, and the NH_4^+ was distilled as NH_3 after addition of base. Gaseous NH_3 was trapped in an acid solution, then trapped onto zeolite and gravity filtered. The zeolite and filter were analyzed as solid samples according to Fry and others (1992). Solid and particulate samples for S isotopic ratios were prepared as SO_2 by use of a sealed-tube combustion technique with KNO_3 as the oxidant (White and others, 1989). The combusted

Table 2. Analytical method references and concentration reporting levels

[mg/L, milligrams per liter; mg/kg, milligrams per kilogram; n.a., not analyzed; ASTM; American Society for Testing and Materials; EPA; U.S. Environmental Protection Agency; "Sum" and "Difference" indicate concentration is computed by sum or difference, respectively, of the associated analytes]

Analyte	Aqueous samples		Solid samples	
	Analytical method[1]	Reporting level	Analytical method[1]	Reporting level
Total carbon	Sum		ASTM D3178	10 mg/kg
Organic carbon	EPA 415.2	0.5 mg/L	Difference	
Inorganic carbon	EPA 415.2	.5 mg/L	ASTM D1756	10 mg/kg
Total nitrogen	Sum		Sum	10 mg/kg
Organic nitrogen	EPA 351.3	.1 mg/L	83-3	10 mg/kg
Ammonia nitrogen	EPA 350.2	.05 mg/L	84-3.5	10 mg/kg
Nitrite nitrogen	EPA 354.1	.05 mg/L	n.a.	10 mg/kg
Nitrate nitrogen	EPA 353.3	.05 mg/L	84-3.5	10 mg/kg
Total sulfur	Sum		Part 2, I-2C	100 mg/kg
Organic sulfur	n.a.		Part 2, I-2C	100 mg/kg
Sulfide as S	EPA 376.2	.02 mg/L	Part 2, I-2C	100 mg/kg
Sulfate as S	EPA 375.4	.3 mg/L	Part 2, I-2C	100 mg/kg

[1]Method references: EPA 350.2, 351.3, 353.3, 354.1, 375.4, 376.2, 415.2 (U.S. Environmental Protection Agency, 1979); ASTM D1756 and D3178 (American Society for Testing and Materials, 1990); 83-3 and 84-3.5 (Black and others, 1965); Part 2, I-2C (Noll and others, 1988).

material was digested in 0.1 N HCl; the resultant solution was filtered through Whatman #4 filters. The filtrate containing dissolved sulfate was heated and $BaCl_2$ was added to precipitate $BaSO_4$. The $BaSO_4$ precipitate was recovered by filtration on Whatman #42 filters and then combusted at 850° C in the presence of V_2O_5 and copper metal according to the method of Yanagisawa and Sakai (1983). Sulfate in water samples was converted to $BaSO_4$ and SO_2 for isotopic analysis in the same manner.

Statistical and Graphical Analytical Methods

Univariate statistical methods, which were used to summarize the data, test for adherence to a normal distribution, and test for differences among data subsets (Mendenhall, 1975; Velleman and Hoaglin, 1981), were done with the computer routines PERCENTILES and EDA (P-STAT, 1989) and UNIVARIATE and GLM (SAS Institute Inc., 1982a, b). Analysis of variance (ANOVA) and multiple comparison tests (MCT) of statistical differences among sample subsets were performed on ranks instead of the original data because the isotopic delta values collected for this study are not normally distributed according to the Shapiro-Wilk statistic (SAS Institute Inc., 1982b). The results of ANOVA and MCT using Tukey's studentized range test (SAS Institute Inc., 1982b) generally are consistent with those of notched boxplots for the same data set, so multiple notched boxplots (P-STAT, 1989) were used as the primary test for differences.

Notched boxplots (for example, see fig. 7) show the distribution of original data values and differences among medians for data subsets or classes relative to a common scale (Velleman and Hoaglin, 1981; Helsel, 1987). The box is defined by the interquartile range (25th to 75th percentiles), and the median (50th percentile) is shown as a "+" within the box. The spread of the two notches "< >" about the median in a boxplot for a class is a function of the variance within the class and defines the 95-percent confidence interval around the median (Velleman and Hoaglin, 1981). If the notched intervals for a pair of boxplots for two classes do not overlap, the medians for the two classes are significantly different at the 95-percent confidence level.

Water-quality data also were evaluated by use of linear and trilinear plots to characterize N-source, soil, and water samples as containing dominantly organic, reduced inorganic, or oxidized inorganic C, N, or S. Concentration data for C (C-org and C-inorg), N (N-org, NH_3-N, and NO_3-N), and S (S-org, H_2S-S,

SO_4-S) are expressed and plotted as percentage of total concentration (for example, see fig. 4). Bivariate plots of isotopic delta values relative to one another and to concentrations were used to examine intrasample and intersample variations. Another bivariate approach involved computation of the molar ratios of total-C to total-N (C:N) and organic-C to total-N (C-org:N) for comparison with data for chemical concentrations and isotope delta values.

INTRASAMPLE VARIATIONS

Field measurements of S.C., pH, temperature, and instantaneous discharge, and laboratory measurements of suspended-sediment (particulate) concentrations in stream-water and selected aqueous samples are given in table B2. The streamflows shown in table B2 generally correlate with the size (drainage area) of the subbasins. Results of the C, N, and S chemical and isotopic analyses are reported in tables B3 and B4, respectively, which list data for individual samples collected in a particular watershed, in chronological order. The format of the Appendix tables B1–B4 facilitates comparison of chemical and isotopic data for samples of various media collected at the same site and time. Before attempting to evaluate isotopic differences between samples or sample groups, however, precision and accuracy of the overall method (sample processing and analysis) must be evaluated. Next, variations within a sample can be assessed by comparison of the data for different sample components, such as the organic fraction relative to the inorganic fraction, the dissolved fraction relative to the particulate fraction, and the dissolved species relative to one another.

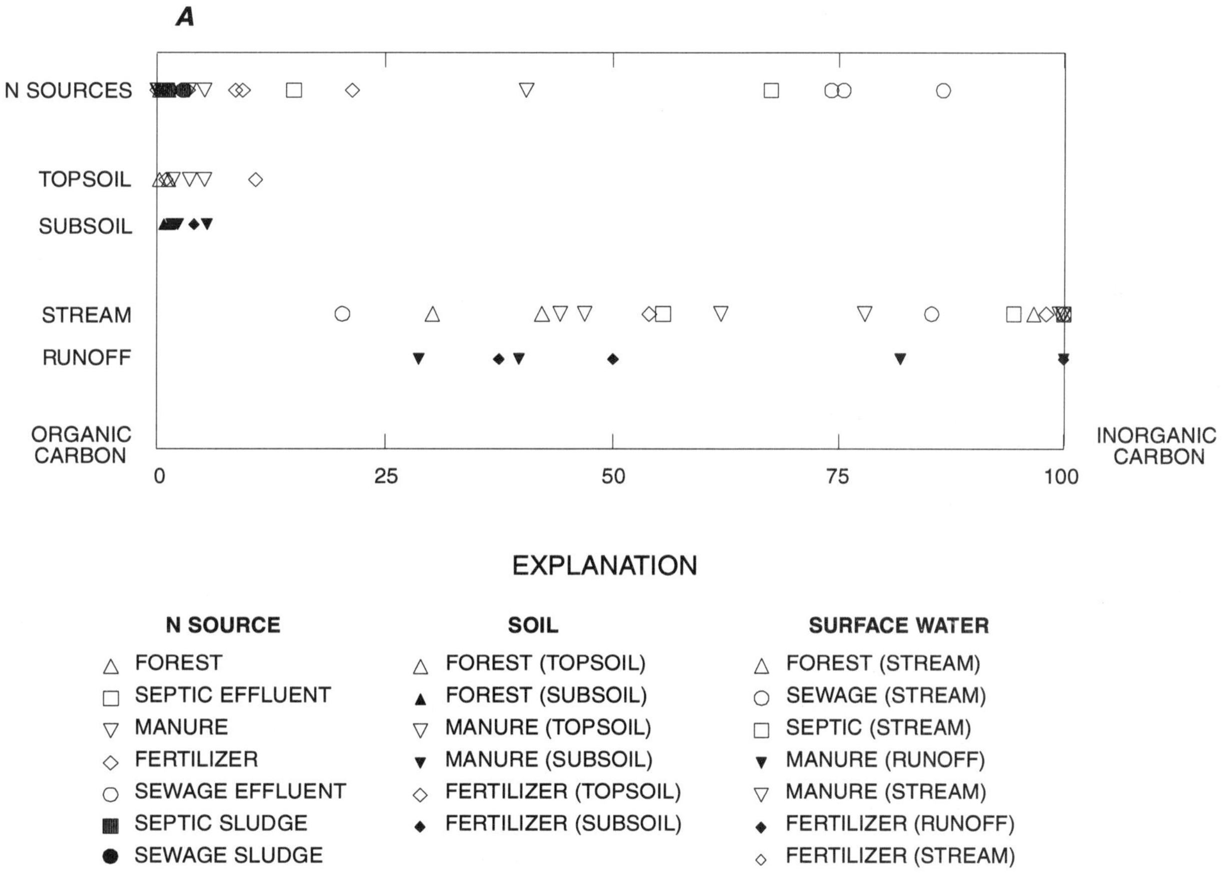

Figure 4. Mean percentage composition of concentrations of organic and inorganic fractions of carbon (C), nitrogen (N), and sulfur (S), relative to concentrations of total C, N, and S, respectively, in N source, soil, and surface-water samples from the Lower Susquehanna River Basin, Pennsylvania: *A*, carbon; *B*, nitrogen; *C*, sulfur.

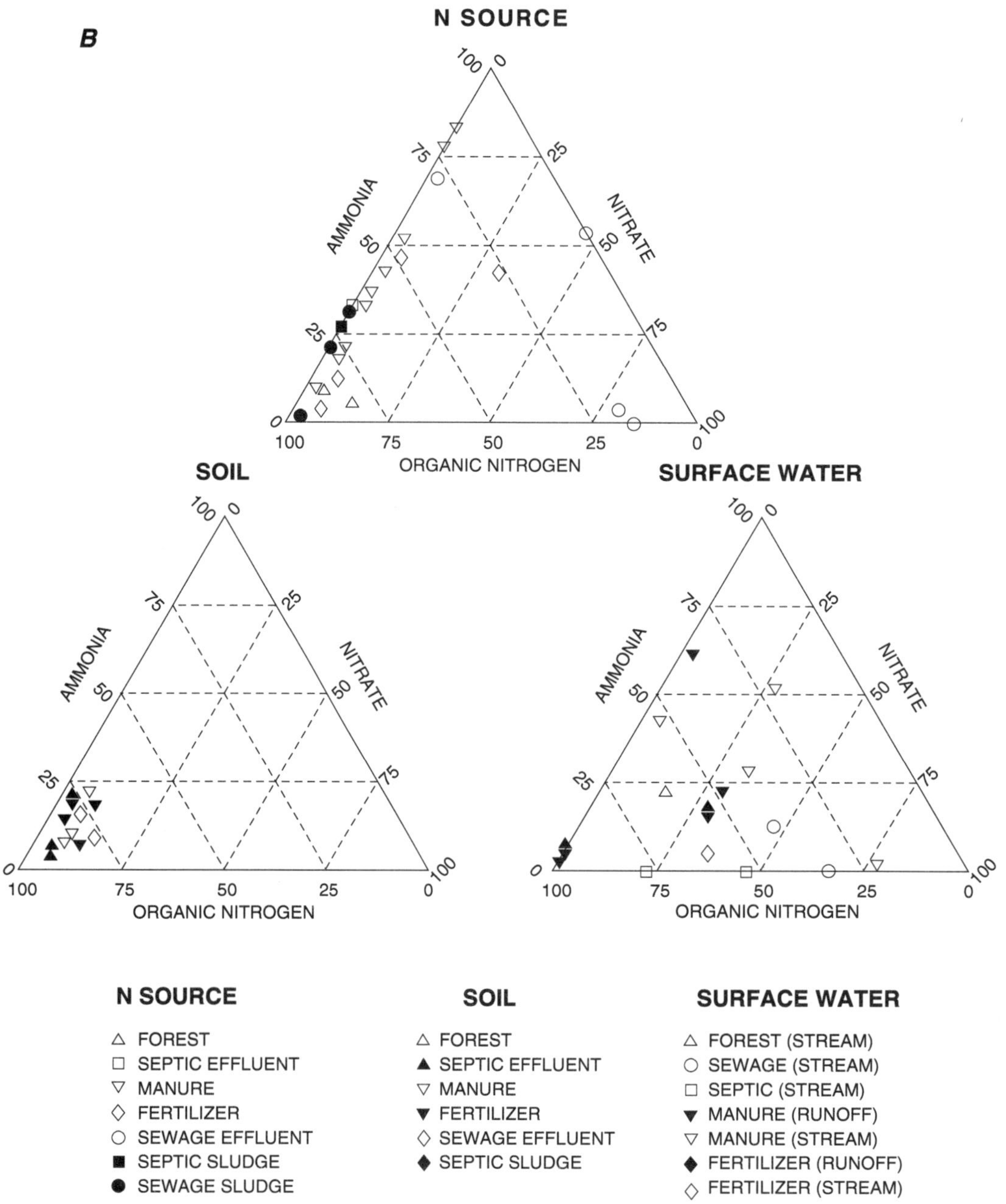

Figure 4. Mean percentage composition of concentrations of organic and inorganic fractions of carbon (C), nitrogen (N), and sulfur (S), relative to concentrations of total C, N, and S, respectively, in N source, soil, and surface-water samples from the Lower Susquehanna River Basin, Pennsylvania: *A*, carbon; *B*, nitrogen; *C*, sulfur—Continued.

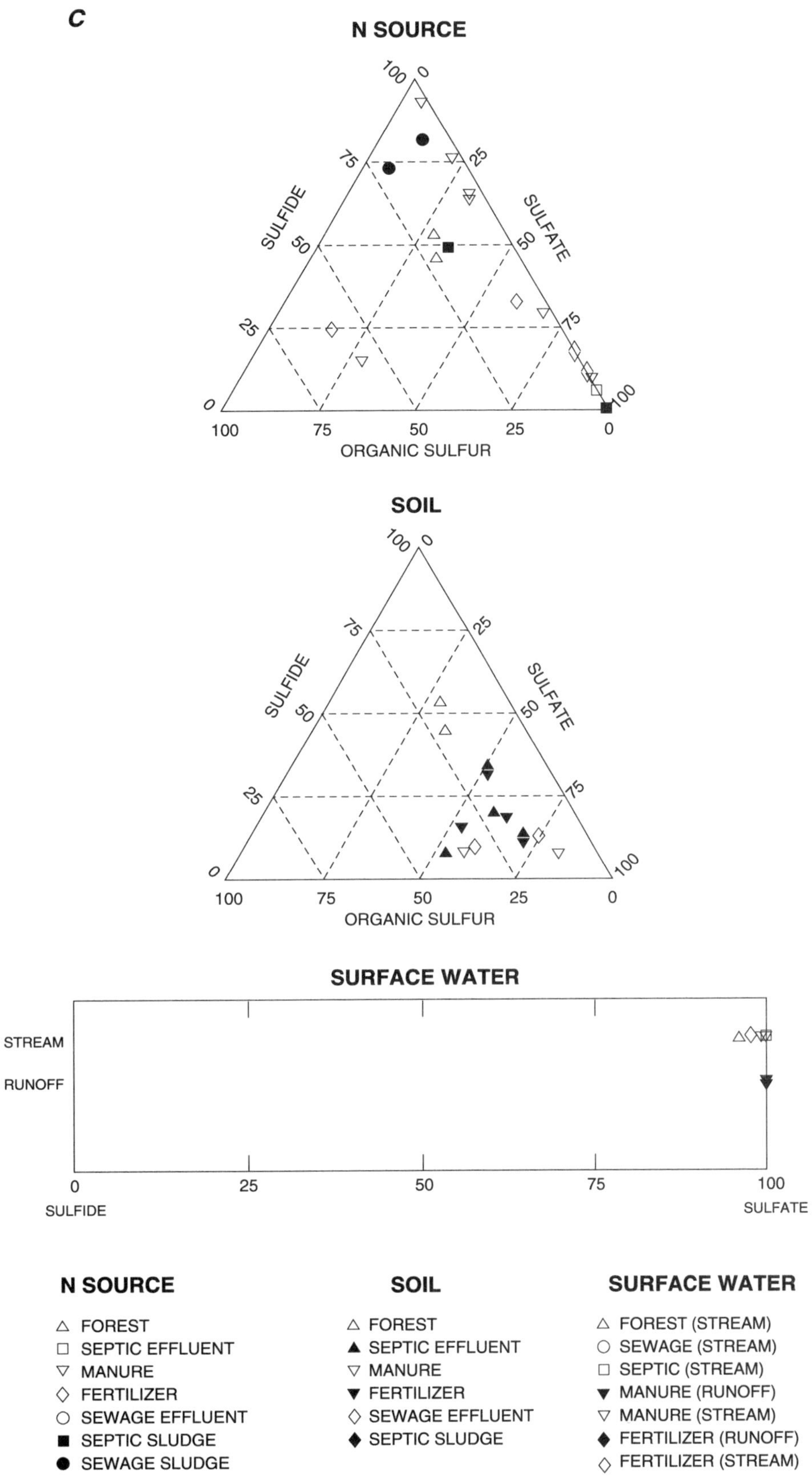

Figure 4. Mean percentage composition of concentrations of organic and inorganic fractions of carbon (C), nitrogen (N), and sulfur (S), relative to concentrations of total C, N, and S, respectively, in N source, soil, and surface-water samples from the Lower Susquehanna River Basin, Pennsylvania: *A*, carbon; *B*, nitrogen; *C*, sulfur—Continued.

Method Precision and Accuracy

Because the primary study objective is to evaluate relative differences in compositions among different samples, overall method precision is essential. A few solid and aqueous samples were split and analyzed in duplicate for concentration, with results that were within about 10 percent. Variations of this magnitude also were apparent in corresponding pairs of whole (unfiltered) and filtered water samples, both of which contained negligible quantities of particulates (suspended sediment less than 1 mg/L) and measurable concentrations of N and S. Filtered samples sometimes were reported to have slightly larger N and S concentrations than the unfiltered samples (table B3). Analysis of filtered and unfiltered blanks and standards indicated that filtration did not produce contamination. Hence, the precision for concentration measurements in stream-water samples is approximately ±0.25 mg/L for N and ±0.5 mg/L for S. Approximate quantities of C and N in samples also were computed by use of an empirical equation that is based on CO_2 or N_2 gas-pressure measurements during isotopic analysis; these computed values generally were comparable with measured concentrations.

Results of duplicate isotopic analyses of split samples reported in table B4 were used to evaluate overall precision. The reported analytical precision for $\delta^{13}C$ and $\delta^{15}N$ measurement of total C and N in homogeneous samples is ±0.1‰ (Fry and others, 1992). However, the overall precision is affected by sample collection, storage, and processing as well as analytical methods. Table 3 summarizes absolute values of the differences between duplicate isotopic measurements for 31 to 40 split samples, where the 100th-percentile value is the largest difference measured. Results of all reported results for duplicate samples, including dissolved and solid samples were combined. On the basis of the 90th-percentile difference, overall precision for $\delta^{13}C$, $\delta^{15}N$, and $\delta^{34}S$ values is about ±0.6‰. The largest differences between duplicate analyses were measured in solid samples and are attributable to inhomogeneities. The small sample size (about 10 µg or less) and analysis of two or more particulate-clogged filters contribute to heterogeneity of solid subsamples. Because isotopic measurements were conducted and expressed relative to the same working standard, accuracy is reasonably assured. Hence, the isotopic method employed herein should enable distinguishing among samples with $\delta^{13}C$, $\delta^{15}N$, or $\delta^{34}S$ values that differ by more than 1.2‰.

Sample Components

Isotopic differences between sample components can be expected because different chemical forms can coexist and can have different tendencies for mobilization, assimilation, or transformation. In this investigation, the isotopic compositions of organic and inorganic fractions of solid and aqueous samples were not analyzed independently; combined fractions were analyzed. Aqueous-sample dissolved inorganic N and S and suspended particulate fractions were analyzed independently; however, no method was available for analysis of the dissolved organic N and S fractions.

Organic and Inorganic

Figure 4 shows that, with the exception of sewage and septic effluents, most N sources and soils contain larger proportions of organic and reduced forms of C, N, and S than inorganic, oxidized forms, whereas most surface water and ground water contain larger proportions of dissolved inorganic C, N, and

Table 3. Summary of differences in results for duplicate isotopic analyses
[Mean and percentile values in per mil]

Isotope	Number of pairs of split samples	Standard deviation	Mean	Percentile				
				50th	75th	90th	95th	100th
^{13}C	31	0.61	0.43	0.20	0.60	1.14	2.34	2.40
^{15}N	40	1.12	.61	.30	.67	1.25	4.68	5.30
^{34}S	31	.44	.44	.30	.50	1.22	1.64	2.00

S species. When organic materials are transformed into inorganic forms, lighter isotopes tend to concentrate in the inorganic products and heavier isotopes tend to concentrate in the residual organic reactant (fig. 3). Additional transformation and fractionation can result during transport, as inorganic species also can be converted into organic compounds within the soil or within the receiving stream by processes of photosynthesis or assimilation (fig. 2). Consequently, the dissolved and particulate fractions in a stream can have different isotopic compositions.

Dissolved and Particulate

Organic-N is the predominant N form in most solid and aqueous samples. However, NH_3-N generally is the predominant inorganic-N species in N sources, with the exception of well-aerated sewage effluents, and NO_3-N generally is the predominant inorganic-N species in surface waters (fig. 4*B*). The $\delta^{15}N$ of dissolved organic N could not be measured. Nevertheless, the $\delta^{15}N$ of particulate materials, which are mostly organic in composition, and dissolved NH_4^+ and NO_3^- could be measured.

Corresponding $\delta^{15}N$-NO_3 and $\delta^{15}N$-NH_3 in aqueous N-source samples are not correlated nor similar (fig. 5), with the exception of liquid-N fertilizer, which is the only N source that has similar $\delta^{15}N$-NO_3 (–2.0‰) and $\delta^{15}N$-NH_3 (–2.4‰). Processes that enrich NH_4^+ in ^{15}N relative to coexisting NO_3^- include volatilization or assimilation of NH_3 and nitrification. A process that enriches NO_3^- in ^{15}N relative to coexisting NH_4^+ is denitrification. The extent to which these processes are completed is key to understanding the isotopic compositions of N sources and associated surface waters.

Because either NH_4^+ or NO_3^- was predominant in most aqueous samples, $\delta^{15}N$-inorg, a mass-weighted average of $\delta^{15}N$-NO_3 and $\delta^{15}N$-NH_3, was computed for comparison with $\delta^{15}N$-org, the corresponding particulate composition. Isotopic differences between the dissolved and particulate fractions in N sources and surface waters are relatively large for N and small for S (table 4, fig. 6); however, the median and mean differences between the fractions for both N and S exceed 1.2‰. Dissolved N and S also have more variable isotopic compositions than the corresponding particulate (fig. 6). The variations in $\delta^{15}N$ and $\delta^{34}S$ in the N-source effluents and surface waters can result from fractionation overprinted on original differences. Additional efforts are made in a later section to estimate the effects of isotopic fractionation on the N isotopic compositions of associated N-source, soil, and surface-water samples.

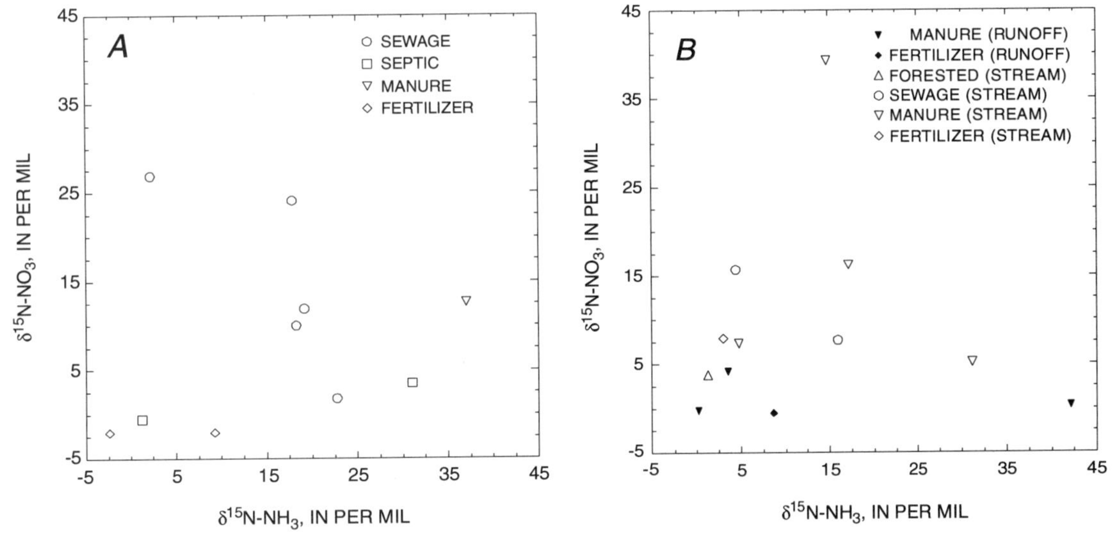

Figure 5. $\delta^{15}N$-NO_3 relative to $\delta^{15}N$-NH_3 of dissolved fractions in aqueous samples, Lower Susquehanna River Basin, Pennsylvania.

Table 4. Summary of differences in isotopic compositions between fractions in aqueous N-source and surface-water samples

Absolute value of difference between[1]	Number of samples	Standard deviation	Mean (per mil)	Percentile (per mil)				
				50th	75th	90th	95th	100th
Aqueous N-source samples								
δ^{15}N-NO_3 – δ^{15}N-NH_3	9	10.78	13.66	8.60	24.55	27.90	27.90	27.90
δ^{15}N-org – δ^{15}N-NO_3	9	7.01	8.85	5.40	15.95	21.10	21.10	21.10
δ^{15}N-org – $\delta^{15}$$NH_3$	9	9.51	18.78	21.30	26.65	30.05	30.05	30.05
δ^{15}N-org – δ^{15}N-inorg	8	10.44	18.75	22.10	27.85	30.04	30.04	30.04
δ^{34}S-org – δ^{34}S-SO_4	13	1.86	2.26	1.70	3.42	5.66	5.90	5.90
Surface-water samples								
δ^{15}N-NO_3 – δ^{15}N-NH_3	15	11.96	9.06	2.65	11.30	32.26	41.80	41.80
δ^{15}N-org – δ^{15}N-NO_3	41	5.46	3.71	2.40	4.55	6.40	11.20	34.30
δ^{15}N-org – δ^{15}N-NH_3	16	8.49	6.96	4.07	7.80	25.54	30.30	30.30
δ^{15}N-org – δ^{15}N-inorg	41	5.74	3.87	2.30	4.55	6.45	22.95	29.82
δ^{34}S-org – δ^{34}S-SO_4	49	1.87	1.97	1.25	3.27	5.00	5.90	7.00

[1] δ^{15}N-NO_3 and δ^{15}N-NH_3 are for dissolved NO_3-N and NH_3-H, respectively; δ^{15}N-inorg is mass-weighted average of δ^{15}N-NO_3 and δ^{15}N-NH_3; δ^{15}N-org is for particulate N. In this table only, δ^{34}S-SO_4 is for dissolved SO_4-S; and δ^{34}S-org is for particulate S; elsewhere, δ^{34}S is used for particulate and dissolved S.

CHARACTERIZATION OF NITROGEN SOURCES AND ASSOCIATED SOILS AND WATERS

Isotopic differences between components of aqueous samples complicate the use of δ^{15}N and δ^{34}S in fluvial samples as tracers of N contamination. Even if particulate and dissolved fractions differ, however, one or the other may have an isotopic composition that is unchanged from that of the N source. Hence, chemical and isotopic compositions of N-source materials and associated soils and waters can be compared with one another to establish if (1) the isotopic compositions of N-source materials differ and (2) the isotopic compositions of nearby soil, runoff-water, and stream-water samples are similar to those of the principal N sources. Tables 5 and 6 summarize the chemical and isotopic data. For the N-source materials sampled, δ^{13}C ranges from about –43 to –12‰; δ^{15}N ranges from about –3.7 to +42‰; δ^{34}S ranges from about –1 to +20‰; and C-org:N ranges from 0.08 to 33. The wide ranges in isotopic and C-org:N ratios indicate potential for different N sources to have distinctive compositions. Chemical and isotopic characteristics of different N sources, and similarities and differences among the N sources and associated soils and waters are described and evaluated in detail below.

Characterization of Nitrogen Sources

Precipitation

Precipitation samples were difficult to analyze for isotopic composition because the water is dilute. Concentrations of total N and total S range from 0.8 to 2.0 mg/L and less than 0.3 to 1.0 mg/L, respectively (table 5, table B3). Precipitation samples collected at manure-use sites (BR1, C1, C2) had the largest concentrations of total N. In all rainwater samples, the largest N –component is organic N (0.3 to 1.2 mg/L); concentrations of NO_3-N (less than 0.1 to 0.3 mg/L) were relatively small. Rain at manure-use sites had concentrations of NH_3-N that were greater than those of NO_3-N; however, rain at synthetic-fertilizer-use sites (BE2, M2A) had concentrations of NO_3-N that were greater than or equal to those of NH_3-N. Because of the acid-buffering effect of dissolved NH_3, the pH of rainwater (table B2) from the manure-use areas (pH 4.6 to 6.5) generally was higher than that from the fertilizer-use areas (pH 4.4 to 4.8). Langland (1992) measured similar effects.

The chemical concentration differences described above indicate potential for N isotopic differences in rainfall of different land-use areas because of elevated NH_3 from animal manure. Despite several attempts to analyze the rain as collected,

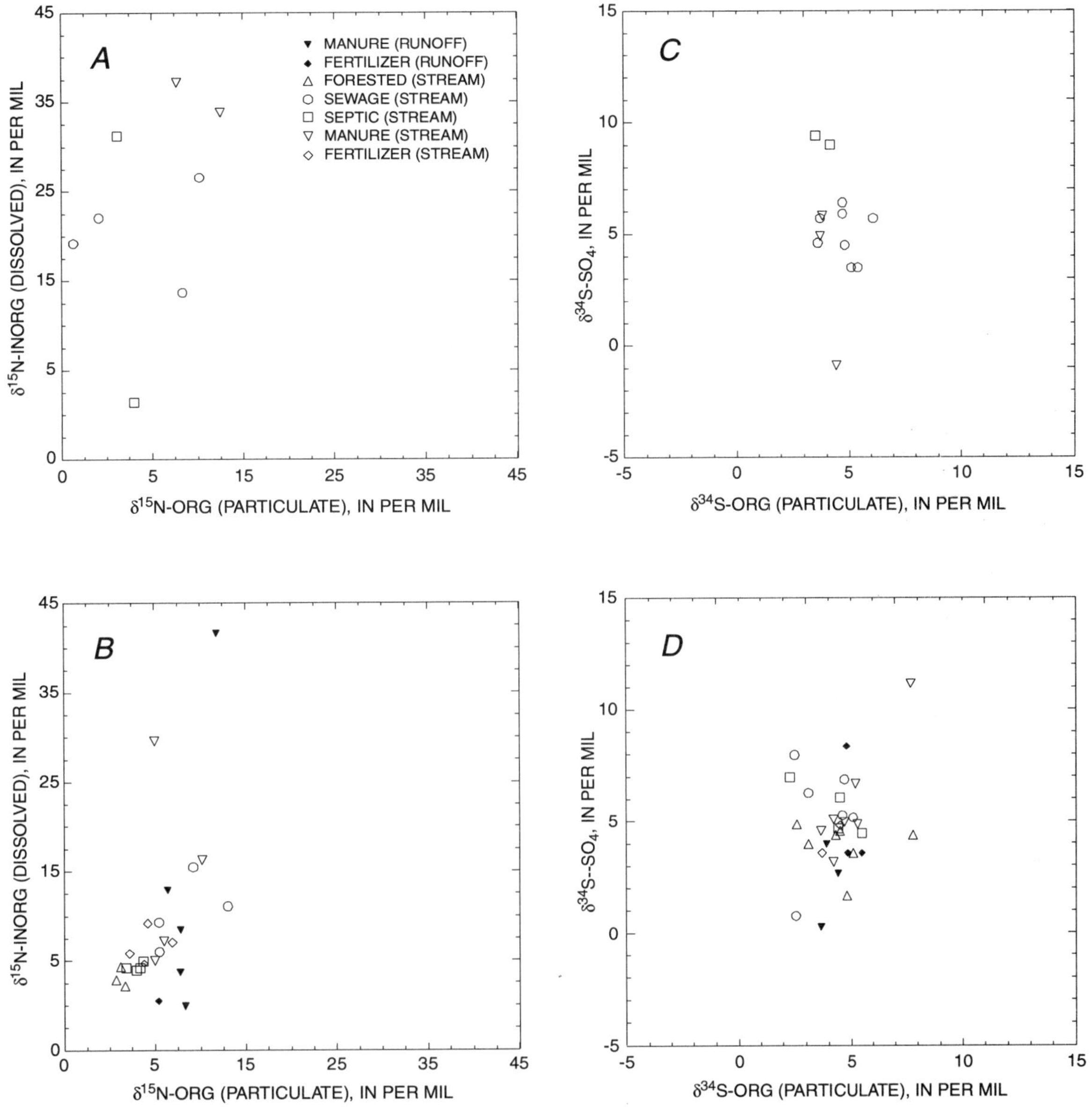

Figure 6. Comparison of dissolved and particulate fractions $\delta^{15}N$ and $\delta^{34}S$ in aqueous samples, Lower Susquehanna River Basin: *A*, $\delta^{15}N$ for N sources; *B*, $\delta^{15}N$ for surface waters; *C*, $\delta^{34}S$ for N sources; *D*, $\delta^{34}S$ for surface waters.

Table 5. Summary statistics for concentrations of carbon, nitrogen, and sulfur in nitrogen sources and associated soil and water in the Lower Susquehanna River Basin, Pennsylvania

[--, not determined for only one sample; <, less than; ppm, concentrations in parts per million]

Sample class	Number of samples	Mean (ppm)	Standard deviation	Coefficient of variation (percent)	Percentile (ppm)				
					100th Maximum	75th	50th Median	25th	0th Minimum
Total carbon (C-total)									
Mixed land use									
Spring water, whole	2	34.75	22.98	66.13	51.0	51.0	34.8	18.5	18.5
Stream water, whole	20	22.83	14.89	65.23	53.6	35.5	21.7	9.5	3.4
Manure land use									
Spring water, whole	2	47.85	21.28	44.48	62.9	62.9	47.9	32.8	32.8
Stream water, whole	9	38.39	21.25	55.36	77.0	54.0	34.0	22.0	7.7
Runoff water, whole	4	168.27	243.75	144.85	533.0	416.8	57.0	31.0	26.0
Subsoil	9	17,255.56	5,162.09	29.92	25,200.0	21,125.0	17,950.0	11,850.0	10,000.0
Topsoil	9	24,505.56	8,526.59	34.79	37,750.0	31,500.0	23,600.0	16,100.0	12,700.0
Steer manure	3	379,616.67	38,619.95	10.17	421,850.0	421,850.0	370,900.0	346,100.0	346,100.0
Cow manure	4	339,575.00	31,330.11	9.23	376,300.0	368,925.0	340,700.0	309,100.0	300,600.0
Pig manure	7	109,799.57	177,588.94	161.74	400,050.0	336,300.0	8,590.0	907.0	600.0
Chicken manure	4	363,437.50	37,224.89	10.24	399,950.0	397,637.5	363,550.0	329,125.0	326,700.0
Fertilizer land use									
Stream water, whole	3	7.97	1.43	17.94	9.2	9.2	8.3	6.4	6.4
Runoff water, whole	4	23.25	3.38	14.53	28.0	26.6	22.5	20.6	20.0
Subsoil	5	18,590.00	3,803.68	20.46	22,600.0	22,200.0	19,250.0	14,650.0	14,600.0
Topsoil	5	36,400.00	26,121.16	71.76	78,600.0	61,450.0	25,400.0	16,850.0	16,300.0
Fertilizer	6	74,349.50	69,146.54	93.00	202,400.0	121,120.2	50,600.0	25,452.5	21,800.0
Sewage land use									
Stream water, whole	6	24.53	6.73	27.45	34.0	30.7	24.5	17.6	16.8
Effluent, whole	7	33.64	15.80	46.96	49.0	42.9	42.0	21.3	4.0
Sludge	2	284,500.00	29,839.91	10.49	305,600.0	305,600.0	284,500.0	263,400.0	263,400.0
Septic land use									
Stream water, whole	6	18.85	18.86	100.06	47.4	40.6	8.2	5.9	4.9
Effluent, whole	3	2,482.20	3,916.68	157.79	7,000.0	7,000.0	403.6	43.0	43.0
Sludge	1	216,500.00	--	--	216,500.0	--	216,500.0	--	216,500.0
Forest land use									
Spring water, whole	1	10.30	--	--	10.3	--	10.3	--	10.3
Stream water, whole	7	5.16	1.87	36.33	8.3	6.1	5.2	3.6	2.4
Subsoil	5	28,660.00	6,983.50	24.37	37,550.0	35,825.0	26,000.0	22,825.0	20,450.0
Topsoil	6	290,300.00	162,464.49	55.96	498,400.0	495,850.0	214,350.0	162,025.0	155,500.0

Table 5. Summary statistics for concentrations of carbon, nitrogen, and sulfur in nitrogen sources and associated soil and water in the Lower Susquehanna River Basin, Pennsylvania—Continued

[--, not determined for only one sample; <, less than; ppm, concentrations in parts per million]

Sample class	Number of samples	Mean (ppm)	Standard deviation	Coefficient of variation (percent)	Percentile (ppm)				
					100th Maximum	75th	50th Median	25th	0th Minimum
Organic carbon (C-org)									
Mixed land use									
Spring water, whole	2	8.25	6.72	81.42	13.0	13.0	8.2	3.5	3.5
Stream water, whole	20	4.93	3.98	80.71	16.0	5.4	4.0	2.0	<1.0
Manure land use									
Spring water, whole	2	<1.00	0	8.32	<1.0	<1.0	<1.0	<1.0	<1.0
Stream water, whole	9	13.88	11.07	79.76	30.0	25.0	11.0	3.5	1.0
Runoff water, whole	4	108.52	181.73	167.46	380.0	295.2	25.5	4.8	.1
Subsoil	9	16,755.56	5,272.36	31.47	24,600.0	20,750.0	17,900.0	11,000.0	9,700.0
Topsoil	9	23,711.11	8,586.10	36.21	37,700.0	30,450.0	23,300.0	15,200.0	12,100.0
Steer manure	3	379,266.67	38,930.37	10.26	421,800.0	421,800.0	370,600.0	345,400.0	345,400.0
Cow manure	4	323,600.00	29,927.36	9.25	363,000.0	354,875.0	316,850.0	299,075.0	297,700.0
Pig manure	7	109,207.57	177,236.87	162.29	400,000.0	333,900.0	8,400.0	870.0	370.0
Chicken manure	4	362,525.00	37,732.07	10.41	399,900.0	397,375.0	362,250.0	327,950.0	325,700.0
Fertilizer land use									
Stream water, whole	3	3.67	1.39	37.82	5.2	5.2	3.3	2.5	2.5
Runoff water, whole	4	13.12	7.19	54.80	20.0	19.5	14.0	5.9	4.5
Subsoil	5	18,040.00	3,621.19	20.07	21,600.0	21,300.0	19,200.0	14,200.0	13,900.0
Topsoil	5	34,880.00	26,626.06	76.34	77,700.0	60,850.0	21,200.0	15,750.0	15,700.0
Fertilizer	6	68,145.00	63,562.81	93.28	184,900.0	116,725.0	42,100.0	23,527.5	21,600.0
Sewage land use									
Stream water, whole	6	8.63	8.03	92.96	22.0	16.7	4.7	2.8	2.4
Effluent, whole	7	6.66	3.85	57.83	12.0	11.0	4.3	3.5	3.0
Sludge	2	281,700.00	31,254.12	11.09	303,800.0	303,800.0	281,700.0	259,600.0	259,600.0
Septic land use									
Stream water, whole	6	2.82	.96	34.23	4.2	3.5	2.8	2.1	1.4
Effluent, whole	3	2,101.33	3,295.11	156.81	5,900.0	5,900.0	390.0	14.0	14.0
Sludge	1	214,200.00	--	--	214,200.0	--	214,200.0	--	214,200.0
Forest land use									
Spring water, whole	1	<1.00	--	--	<1.0	--	<1.0	--	<1.0
Stream water, whole	7	3.41	1.74	50.85	6.3	4.2	3.6	1.5	1.1
Subsoil	5	28,320.00	7,035.77	24.84	37,500.0	35,600.0	25,100.0	22,650.0	20,400.0
Topsoil	6	288,350.00	162,304.87	56.29	497,100.0	493,950.0	210,900.0	161,175.0	155,100.0

Table 5. Summary statistics for concentrations of carbon, nitrogen, and sulfur in nitrogen sources and associated soil and water in the Lower Susquehanna River Basin, Pennsylvania—Continued

[--, not determined for only one sample; <, less than; ppm, concentrations in parts per million]

Sample class	Number of samples	Mean (ppm)	Standard deviation	Coefficient of variation (percent)	Percentile (ppm)				
					100th Maximum	75th	50th Median	25th	0th Minimum
Inorganic carbon (C-inorg)									
Mixed land use									
Spring water, whole	2	26.50	16.26	61.37	38.0	38.0	26.5	15.0	15.0
Stream water, whole	20	17.90	15.47	86.45	52.0	32.5	11.5	5.0	<5.0
Manure land use									
Spring water, whole	2	47.00	21.21	45.13	62.0	62.0	47.0	32.0	32.0
Stream water, whole	9	24.51	20.06	81.86	66.0	36.5	20.0	9.4	<5.0
Runoff water, whole	4	59.75	63.15	105.69	153.0	125.5	35.0	18.7	16.0
Subsoil	9	500.00	323.07	64.61	1,000.0	700.0	600.0	100.0	<100.0
Topsoil	9	794.44	483.76	60.89	1,500.0	1,250.0	600.0	450.0	<100.0
Steer manure	3	350.00	327.87	93.68	700.0	700.0	300.0	<100.0	<100.0
Cow manure	4	15,975.00	11,777.77	73.73	31,400.0	27,625.0	14,800.0	5,500.0	2,900.0
Pig manure	7	592.00	898.24	151.73	2,400.0	1,200.0	190.0	37.0	<100.0
Chicken manure	4	912.50	676.23	74.11	1,700.0	1,525.0	950.0	150.0	<100.0
Fertilizer land use									
Stream water, whole	3	4.30	.61	14.15	5.0	5.0	<5.0	<5.0	<5.0
Runoff water, whole	4	10.12	5.72	56.50	18.0	16.0	9.0	5.4	<5.0
Subsoil	5	550.00	406.20	73.85	1,000.0	900.0	700.0	<100.0	<100.0
Topsoil	5	1,520.00	1,573.85	103.54	4,200.0	2,900.0	900.0	450.0	300.0
Fertilizer	6	6,204.50	8,050.66	129.76	17,500.0	16,000.0	2,000.0	156.7	27.0
Sewage land use									
Stream water, whole	6	15.90	10.64	66.92	31.0	25.7	15.5	6.1	1.8
Effluent, whole	7	26.99	14.29	52.94	39.0	38.0	34.0	17.4	<1.0
Sludge	2	2,800.00	1,414.21	50.51	3,800.0	3,800.0	2,800.0	1,800.0	1,800.0
Septic land use									
Stream water, whole	6	16.03	19.34	120.61	46.0	37.7	5.3	<5.0	<5.0
Effluent, whole	3	380.87	622.84	163.53	1,100.0	1,100.0	29.0	13.6	13.6
Sludge	1	2,300.00	--	--	2,300.0	--	2,300.0	--	2,300.0
Forest land use									
Spring water, whole	1	9.80	--	--	9.8	--	9.8	--	9.8
Stream water, whole	7	<5.00	--	--	<5.0	<5.0	<5.0	<5.0	<5.0
Subsoil	5	340.00	348.93	102.63	900.0	650.0	300.0	<100.0	<100.0
Topsoil	6	1,950.00	2,317.54	118.85	6,500.0	3,200.0	1,150.0	400.0	400.0

Table 5. Summary statistics for concentrations of carbon, nitrogen, and sulfur in nitrogen sources and associated soil and water in the Lower Susquehanna River Basin, Pennsylvania—Continued

[--, not determined for only one sample; <, less than; ppm, concentrations in parts per million]

Sample class	Number of samples	Mean (ppm)	Standard deviation	Coefficient of variation (percent)	Percentile (ppm)				
					100th Maximum	75th	50th Median	25th	0th Minimum
Total nitrogen (N-total)									
Mixed land use									
Spring water, dissolved	2	1.05	0.14	13.47	1.2	1.2	1.1	1.0	1.0
Spring water, whole	2	1.67	.88	52.77	2.3	2.3	1.7	1.0	1.0
Stream water, dissolved	20	2.70	3.27	120.74	16.0	2.9	1.9	1.2	.8
Stream water, whole	20	2.70	2.70	100.04	13.5	2.8	2.0	1.5	1.0
Manure land use									
Spring water, dissolved	2	10.02	9.44	94.16	16.7	16.7	10.0	3.3	3.3
Spring water, whole	2	9.57	8.38	87.51	15.5	15.5	9.6	3.6	3.6
Stream water, dissolved	8	6.29	2.18	34.69	8.6	8.1	6.6	4.9	2.1
Stream water, whole	8	6.50	2.94	45.23	9.6	9.3	6.5	3.6	1.9
Runoff water, dissolved	4	23.94	43.11	180.09	88.6	67.2	2.5	2.1	2.1
Runoff water, whole	4	32.54	59.65	183.32	122.0	92.4	3.2	2.0	1.7
Subsoil	9	1,913.00	567.24	29.65	2,880.0	2,404.5	1,855.0	1,390.0	1,180.0
Topsoil	9	2,405.33	943.74	39.24	4,260.0	2,990.0	2,296.0	1,515.0	1,260.0
Steer manure	3	32,171.67	2,822.03	8.77	34,405.0	34,405.0	33,110.0	29,000.0	29,000.0
Cow manure	4	25,247.50	4,792.71	18.98	31,213.0	29,797.2	25,135.0	20,810.2	19,507.0
Pig manure	12	7,467.03	11,718.77	156.94	34,350.0	4,455.7	3,052.8	1,286.7	793.1
Chicken manure	4	62,932.50	19,979.79	31.75	80,420.0	80,322.5	63,250.0	45,225.0	44,810.0
Fertilizer land use									
Stream water, dissolved	3	1.40	.54	38.63	2.0	2.0	1.2	.9	.9
Stream water, whole	3	1.87	.43	22.78	2.3	2.3	1.8	1.4	1.4
Runoff water, dissolved	4	1.70	1.13	66.59	3.3	2.9	1.3	.9	.8
Runoff water, whole	4	5.94	5.94	99.99	13.3	12.0	4.8	1.0	.9
Subsoil	5	1,943.40	443.68	22.83	2,540.0	2,420.0	1,650.0	1,613.5	1,597.0
Topsoil	5	3,373.40	1,752.26	51.94	6,240.0	4,910.0	2,990.0	2,028.5	1,670.0
Fertilizer	6	204,939.17	114,043.28	55.65	353,000.0	343,400.0	162,890.0	115,563.7	84,855.0
Sewage land use									
Stream water, dissolved	6	3.66	1.53	41.73	5.8	5.1	3.6	2.0	1.9
Stream water, whole	6	3.83	1.54	40.27	6.0	5.3	3.8	2.4	1.8
Effluent, dissolved	7	8.24	5.63	68.29	19.9	9.4	5.8	5.3	2.6
Effluent, whole	7	10.47	7.27	69.40	24.2	16.0	8.7	5.9	2.9
Sludge	2	38,690.00	10,465.18	27.05	46,090.0	46,090.0	38,690.0	31,290.0	31,290.0
Septic land use									
Stream water, dissolved	6	.97	.48	49.58	1.6	1.5	.8	.6	.6
Stream water, whole	6	1.29	.85	65.80	2.8	1.9	1.0	.7	.5
Effluent, dissolved	3	77.07	50.58	65.63	110.2	110.2	102.1	18.8	18.8
Effluent, whole	3	293.30	338.76	115.50	672.0	672.0	188.6	19.2	19.2
Sludge	1	33,280.00	--	--	33,280.0	--	33,280.0	--	33,280.0
Forest land use									
Spring water, dissolved	1	.55	--	--	.6	--	.6	--	.6
Spring water, whole	1	.55	--	--	.6	--	.6	--	.6
Stream water, dissolved	7	.94	.45	47.67	1.9	1.0	.8	.6	.6
Stream water, whole	7	.94	.52	54.98	2.1	.8	.8	.6	.6
Subsoil	5	1,367.40	88.73	6.49	1,484.0	1,442.0	1,370.0	1,291.5	1,240.0
Topsoil	6	11,903.17	7,189.48	60.40	21,488.0	20,822.0	8,215.5	6,555.0	6,240.0
Precipitation	7	1.20	.43	36.08	2.0	1.4	1.0	.8	.8

Table 5. Summary statistics for concentrations of carbon, nitrogen, and sulfur in nitrogen sources and associated soil and water in the Lower Susquehanna River Basin, Pennsylvania—Continued

[--, not determined for only one sample; <, less than; ppm, concentrations in parts per million]

Sample class	Number of samples	Mean (ppm)	Standard deviation	Coefficient of variation (percent)	Percentile (ppm)				
					100th Maximum	75th	50th Median	25th	0th Minimum
Organic nitrogen (N-org)									
Mixed land use									
Spring water, dissolved	2	0.65	0.21	32.64	0.8	0.8	0.6	0.5	0.5
Spring water, whole	2	.90	.42	47.14	1.2	1.2	.9	.6	.6
Stream water, dissolved	20	.76	.44	57.81	2.4	.9	.6	.5	.4
Stream water, whole	20	.84	.43	50.72	2.2	1.1	.6	.5	.5
Manure land use									
Spring water, dissolved	2	.40	.14	35.36	.5	.5	.4	.3	.3
Spring water, whole	2	.35	.07	20.20	.4	.4	.3	.3	.3
Stream water, dissolved	9	1.48	1.54	104.44	4.1	3.1	.7	.4	.2
Stream water, whole	9	1.79	1.53	85.49	4.8	3.0	1.2	.5	.4
Runoff water, dissolved	4	9.90	17.55	177.26	36.2	27.6	1.6	.4	.2
Runoff water, whole	4	12.95	22.20	171.40	46.2	35.5	2.5	.8	.5
Subsoil	9	1,472.33	556.23	37.78	2,460.0	1,884.5	1,410.0	1,005.0	760.0
Topsoil	9	1,861.11	976.83	52.49	3,753.0	2,400.0	1,790.0	965.0	550.0
Steer manure	3	20,870.67	10,669.07	51.12	29,552.0	29,552.0	24,100.0	8,960.0	8,960.0
Cow manure	4	14,435.50	5,879.77	40.73	19,248.0	18,753.5	16,267.0	8,286.0	5,960.0
Pig manure	12	3,618.42	7,544.50	208.50	23,410.0	1,485.0	350.0	133.7	16.0
Chicken manure	4	48,686.00	13,348.50	27.42	65,954.0	62,568.0	45,820.0	37,670.0	37,150.0
Fertilizer land use									
Stream water, dissolved	3	.80	.52	64.95	1.4	1.4	.5	.5	.5
Stream water, whole	3	1.10	.40	36.36	1.5	1.5	1.1	.7	.7
Runoff water, dissolved	4	.95	.40	42.54	1.5	1.4	.8	.6	.6
Runoff water, whole	4	3.42	4.72	137.82	10.5	8.2	1.2	.9	.8
Subsoil	5	1,514.40	316.87	20.92	1,930.0	1,845.0	1,392.0	1,245.0	1,180.0
Topsoil	5	2,591.00	1,479.59	57.10	5,140.0	3,815.0	2,105.0	1,610.0	1,400.0
Fertilizer	6	99,371.67	111,736.85	112.44	300,700.0	184,075.0	65,950.0	13,472.5	700.0
Sewage land use									
Stream water, dissolved	6	1.12	.75	66.87	2.4	1.8	.8	.6	.4
Stream water, whole	6	1.42	.88	61.93	2.9	2.2	1.1	.7	.7
Effluent, dissolved	7	.71	.37	52.03	1.2	1.1	.7	.5	.1
Effluent, whole	7	2.30	2.77	120.54	8.4	2.6	1.4	.7	.7
Sludge	2	28,100.00	5,388.15	19.17	31,910.0	31,910.0	28,100.0	24,290.0	24,290.0
Septic land use									
Stream water, dissolved	6	.52	.12	22.63	.7	.6	.5	.4	.4
Stream water, whole	6	.78	.42	53.81	1.6	1.0	.6	.5	.4
Effluent, dissolved	3	1.08	.98	90.48	2.0	2.0	1.2	<.1	<.1
Effluent, whole	3	195.23	316.11	161.91	560.0	560.0	24.4	1.3	1.3
Sludge	1	24,270.00	--	--	24,270.0	--	24,270.0	--	24,270.0
Forest land use									
Spring water, dissolved	1	.40	--	--	.4	--	.4	--	.4
Spring water, whole	1	.40	--	--	.4	--	.4	--	.4
Stream water, dissolved	7	.71	.23	32.76	1.1	.9	.7	.5	.5
Stream water, whole	7	.69	.25	36.14	1.2	.7	.7	.5	.5
Subsoil	5	1,156.20	270.13	23.36	1,412.0	1,376.0	1,189.0	920.0	720.0
Topsoil	6	9,833.83	5,790.91	58.89	19,678.0	15,344.5	7,167.5	5,687.5	5,260.0
Precipitation	7	.66	.42	63.27	1.2	1.2	.6	.3	.1

Table 5. Summary statistics for concentrations of carbon, nitrogen, and sulfur in nitrogen sources and associated soil and water in the Lower Susquehanna River Basin, Pennsylvania—Continued

[--, not determined for only one sample; <, less than; ppm, concentrations in parts per million]

Sample class	Number of samples	Mean (ppm)	Standard deviation	Coefficient of variation (percent)	Percentile (ppm)				
					100th Maximum	75th	50th Median	25th	0th Minimum
Inorganic nitrogen (N-inorg)									
Mixed land use									
Spring water, dissolved	2	0.40	0.35	88.39	0.7	0.7	0.4	0.2	<0.2
Spring water, whole	2	.77	.46	59.31	1.1	1.1	.8	.4	.5
Stream water, dissolved	20	1.94	3.28	168.78	15.5	2.0	1.3	.6	.4
Stream water, whole	20	1.85	2.72	147.08	13.0	1.8	1.4	.6	.4
Manure land use									
Spring water, dissolved	2	9.62	9.30	96.61	16.2	16.2	9.6	3.0	3.0
Spring water, whole	2	9.22	8.31	90.07	15.1	15.1	9.2	3.4	3.4
Stream water, dissolved	8	4.70	1.94	41.22	8.4	5.2	4.6	4.0	1.4
Stream water, whole	8	4.64	2.58	55.68	9.2	5.8	4.8	2.1	1.2
Runoff water, dissolved	4	14.04	25.60	182.35	52.4	40.0	1.8	.3	.2
Runoff water, whole	4	19.59	37.48	191.34	75.8	57.2	1.2	.4	.2
Subsoil	9	440.67	209.38	47.52	890.0	575.0	390.0	289.5	220.0
Topsoil	9	544.22	180.25	33.12	820.0	700.0	510.0	445.5	230.0
Steer manure	3	11,301.00	7,848.42	69.45	20,040.0	20,040.0	9,010.0	4,853.0	4,853.0
Cow manure	4	10,812.00	6,654.26	61.55	19,590.0	17,683.8	9,707.5	5,044.8	4,243.0
Pig manure	12	3,848.62	5,145.41	133.69	19,180.0	4,025.6	2,143.6	1,125.7	777.1
Chicken manure	4	14,246.50	9,631.15	67.60	27,620.0	24,331.5	11,893.0	6,515.0	5,580.0
Fertilizer land use									
Stream water, dissolved	3	.60	.15	25.00	.8	.8	.6	.4	.4
Stream water, whole	3	.77	.03	3.77	.8	.8	.8	.8	.8
Runoff water, dissolved	4	.75	.82	109.27	1.9	1.6	.5	.1	.1
Runoff water, whole	4	2.51	3.25	129.22	7.0	6.0	1.5	.1	.1
Subsoil	5	429.00	164.94	38.45	610.0	575.0	470.0	262.5	205.0
Topsoil	5	782.40	463.32	59.22	1,170.0	1,135.0	1,090.0	276.0	270.0
Fertilizer	6	105,567.50	88,257.15	83.60	263,000.0	161,712.5	84,027.5	38,350.0	34,900.0
Sewage land use									
Stream water, dissolved	6	2.54	1.26	49.59	4.5	3.7	2.4	1.4	1.2
Stream water, whole	6	2.42	1.21	50.13	4.4	3.4	2.3	1.2	1.2
Effluent, dissolved	7	7.53	5.83	77.44	19.8	8.2	5.2	4.2	2.2
Effluent, whole	7	8.17	4.84	59.29	15.8	13.4	7.3	5.0	2.2
Sludge	2	10,590.00	5,077.03	47.94	14,180.0	14,180.0	10,290.0	7,000.0	7,000.0
Septic land use									
Stream water, dissolved	6	.45	.37	81.95	1.0	.9	.3	.2	.2
Stream water, whole	6	.51	.45	89.38	1.2	1.0	.4	.1	.1
Effluent, dissolved	3	75.98	51.33	67.55	109.0	109.0	102.0	16.8	16.8
Effluent, whole	3	98.07	74.12	75.58	164.2	164.2	112.0	17.9	18.0
Sludge	1	9,010.00	--	--	9,010.0	--	9,010.0	--	9,010.0
Forest land use.									
Spring water, dissolved	1	.20	--	--	.2	--	.2	--	.2
Spring water, whole	1	.20	--	--	.2	--	.2	--	.2
Stream water, dissolved	7	.23	.26	112.86	.8	.2	.1	.1	.1
Stream water, whole	7	.26	.30	115.81	.9	.4	.1	.1	.1
Subsoil	5	211.20	188.66	89.33	520.0	385.0	154.0	66.0	60.0
Topsoil	6	2,069.33	2,298.07	111.05	6,700.0	3,032.5	1,118.0	837.5	830.0
Precipitation	7	.54	.57	105.34	1.7	.9	.2	.2	.2

Table 5. Summary statistics for concentrations of carbon, nitrogen, and sulfur in nitrogen sources and associated soil and water in the Lower Susquehanna River Basin, Pennsylvania—Continued

[--, not determined for only one sample; <, less than; ppm, concentrations in parts per million]

Sample class	Number of samples	Mean (ppm)	Standard deviation	Coefficient of variation (percent)	Percentile (ppm)				
					100th Maximum	75th	50th Median	25th	0th Minimum
Nitrate nitrogen (NO_3-N)									
Mixed land use									
Spring water, dissolved	2	0.32	0.39	119.66	0.6	0.6	0.2	<0.1	<0.1
Spring water, whole	2	.70	.42	60.61	1.0	1.0	.7	.4	.4
Stream water, dissolved	20	1.13	.77	67.54	3.1	1.6	.9	.5	.3
Stream water, whole	20	1.12	.73	64.80	3.1	1.5	1.0	.5	.3
Manure land use									
Spring water, dissolved	2	9.50	9.19	96.76	16.0	16.0	9.5	3.0	3.0
Spring water, whole	2	9.15	8.27	90.42	15.0	15.0	9.1	3.3	3.3
Stream water, dissolved	8	3.12	2.07	66.20	5.2	4.8	4.0	.8	.2
Stream water, whole	8	2.65	2.16	81.41	5.6	5.0	1.7	.7	.5
Runoff water, dissolved	4	.74	.75	101.37	1.8	1.5	.5	.0	<.1
Runoff water, whole	4	.84	.83	98.63	2.0	1.7	.6	.1	<.1
Subsoil	9	128.33	73.52	57.29	290.0	165.0	111.0	84.5	50.0
Topsoil	9	159.89	90.30	56.47	290.0	250.0	123.0	75.0	60.0
Steer manure	3	875.00	336.34	38.44	1,250.0	1,250.0	775.0	600.0	600.0
Cow manure	4	743.75	213.43	28.70	1,060.0	965.2	657.5	608.5	600.0
Pig manure	12	186.45	433.80	232.66	1,230.0	5.7	1.1	.7	.5
Chicken manure	4	1,659.25	1,050.84	63.33	3,160.0	2,759.2	1,353.5	865.0	770.0
Fertilizer land use									
Stream water, dissolved	3	.53	.15	28.64	.7	.7	.5	.4	.4
Stream water, whole	3	.67	.06	8.66	.7	.7	.7	.6	.6
Runoff water, dissolved	4	.67	.83	122.84	1.8	1.5	.3	<.1	<.1
Runoff water, whole	4	1.52	1.96	128.85	4.2	3.6	.8	<.1	<.1
Subsoil	5	94.60	52.59	55.59	180.0	145.0	70.0	56.5	53.0
Topsoil	5	178.20	131.29	73.67	390.0	305.0	111.0	85.0	70.0
Fertilizer	6	24,474.17	42,829.53	175.00	110,000.0	45,650.0	5,932.5	445.0	280.0
Sewage land use									
Stream water, dissolved	6	2.23	1.19	53.33	4.4	3.0	2.0	1.2	1.1
Stream water, whole	6	2.18	1.17	53.37	4.3	2.9	2.0	1.2	1.1
Effluent, dissolved	7	3.59	2.82	78.54	8.2	5.2	3.9	.8	.1
Effluent, whole	7	4.03	3.15	78.14	8.2	7.0	5.0	.6	.1
Sludge	2	60.00	70.71	117.85	110.0	110.0	60.0	10.0	10.0
Septic land use									
Stream water, dissolved	6	.40	.37	92.20	.9	.8	.2	.1	.1
Stream water, whole	6	.46	.45	99.13	1.2	.9	.3	.0	<.1
Effluent, dissolved	3	<.10	--	--	<.1	<.1	<.1	<.1	<.1
Effluent, whole	3	.10	.09	86.60	.2	.2	<.1	<.1	<.1
Sludge	1	10.00	--	--	10.0	--	10.0	--	10.0
Forest land use									
Spring water, dissolved	1	.10	--	--	.1	--	.1	--	.1
Spring water, whole	1	.10	--	--	.1	--	.1	--	.1
Stream water, dissolved	7	.11	.17	148.82	.5	<.1	<.1	<.1	<.1
Stream water, whole	7	.11	.17	148.82	.5	<.1	<.1	<.1	<.1
Subsoil	5	64.20	38.78	60.41	120.0	98.5	60.0	32.0	14.0
Topsoil	6	1,084.50	2,018.59	186.13	5,200.0	1,619.5	280.5	165.0	150.0
Precipitation	7	.17	.12	71.16	.3	.3	.1	<.1	<.1

Table 5. Summary statistics for concentrations of carbon, nitrogen, and sulfur in nitrogen sources and associated soil and water in the Lower Susquehanna River Basin, Pennsylvania—Continued

[--, not determined for only one sample; <, less than; ppm, concentrations in parts per million]

Sample class	Number of samples	Mean (ppm)	Standard deviation	Coefficient of variation (percent)	Percentile (ppm)				
					100th Maximum	75th	50th Median	25th	0th Minimum
Ammonia nitrogen (NH_3-N)									
Mixed land use									
Spring water, dissolved	2	0.07	0.04	47.14	0.1	0.1	<0.1	<0.1	<0.1
Spring water, whole	2	.07	.04	47.14	.1	.1	<.1	<.1	<.1
Stream water, dissolved	20	.81	3.11	383.66	14.0	.1	<0.1	<.1	<.1
Stream water, whole	20	.73	2.66	365.72	12.0	.2	<.1	<.1	<.1
Manure land use									
Spring water, dissolved	2	.12	.11	84.85	.2	.2	<.1	<.1	<.1
Spring water, whole	2	.07	.04	47.14	.1	.1	<.1	<.1	<.1
Stream water, dissolved	9	1.54	1.91	123.95	4.8	3.6	.3	<.1	<.1
Stream water, whole	9	1.90	2.09	109.83	5.4	4.0	1.2	.1	.1
Runoff water, dissolved	4	13.30	25.67	193.00	51.8	39.1	.6	.2	.1
Runoff water, whole	4	18.75	36.70	195.74	73.8	55.5	.5	.2	.1
Subsoil	9	312.33	209.90	67.20	800.0	395.0	202.0	183.5	170.0
Topsoil	9	384.33	169.44	44.09	760.0	465.0	350.0	281.0	170.0
Steer manure	3	10,426.00	8,020.50	76.93	19,440.0	19,440.0	7,760.0	4,078.0	4,078.0
Cow manure	4	10,068.25	6,741.10	66.95	18,990.0	17,063.5	8,837.0	4,304.2	3,609.0
Pig manure	12	3,662.17	4,764.98	130.11	17,950.0	4,025.0	2,140.0	1,125.0	776.0
Chicken manure	4	12,587.25	10,187.90	80.94	26,850.0	23,364.7	9,534.5	4,862.5	4,430.0
Fertilizer land use									
Stream water, dissolved	3	.07	.03	43.30	.1	.1	<.1	<.1	<.1
Stream water, whole	3	.10	.09	86.60	.2	.2	<.1	<.1	<.1
Runoff water, dissolved	4	.07	.03	38.49	.1	.1	<.1	<.1	<.1
Runoff water, whole	4	.99	1.28	130.10	2.8	2.3	.5	<.1	<.1
Subsoil	5	334.40	150.82	45.10	500.0	490.0	290.0	201.0	152.0
Topsoil	5	604.20	389.50	64.47	990.0	935.0	780.0	185.5	171.0
Fertilizer	6	81,093.33	60,440.39	74.53	153,000.0	134,002.5	81,895.0	23,925.0	15,300.0
Sewage land use									
Stream water, dissolved	6	.31	.54	173.57	1.4	.4	.1	<.1	<.1
Stream water, whole	6	.23	.43	182.16	1.1	.3	<.1	<.1	<.1
Effluent, dissolved	7	3.94	7.28	184.61	19.0	8.1	<.1	<.1	<.1
Effluent, whole	7	4.14	6.93	167.20	15.2	13.3	<.1	<.1	<.1
Sludge	2	10,530.00	5,147.74	48.89	14,170.0	14,170.0	10,530.0	6,890.0	6,890.0
Septic land use									
Stream water, dissolved	6	<.10	--	--	<.1	<.1	<.1	<.1	<.1
Stream water, whole	6	<.10	--	--	<.1	<.1	<.1	<.1	<.1
Effluent, dissolved	3	75.93	51.33	67.60	109.0	109.0	102.0	16.8	16.8
Effluent, whole	3	97.97	74.05	75.59	164.0	164.0	112.0	17.9	17.9
Sludge	1	9,000.00	--	--	9,000.0	--	9,000.0	-	9,000.0
Forest land use									
Spring water, dissolved	1	<.10	--	--	<.1	--	<.1	--	<.1
Spring water, whole	1	<.10	--	--	<.1	--	<.1	--	<.1
Stream water, dissolved	7	.11	.10	86.45	.3	.2	<.1	<.1	<.1
Stream water, whole	7	.14	.15	101.90	.4	.3	<.1	<.1	<.1
Subsoil	5	147.00	156.08	106.18	400.0	295.0	77.0	34.0	10.0
Topsoil	6	984.83	375.15	38.09	1,500.0	1,413.0	837.5	677.5	670.0
Precipitation	7	.37	.49	131.63	1.4	.6	.1	.1	.1

Table 5. Summary statistics for concentrations of carbon, nitrogen, and sulfur in nitrogen sources and associated soil and water in the Lower Susquehanna River Basin, Pennsylvania—Continued

[--, not determined for only one sample; <, less than; ppm, concentrations in parts per million]

Sample class	Number of samples	Mean (ppm)	Standard deviation	Coefficient of variation (percent)	Percentile (ppm)				
					100th Maximum	75th	50th Median	25th	0th Minimum
Total sulfur (S-total)									
Mixed land use									
Spring water, dissolved	2	3.75	1.41	37.71	4.8	4.8	3.8	2.8	2.8
Spring water, whole	2	2.40	.49	20.62	2.8	2.8	2.4	2.0	2.0
Stream water, dissolved	20	7.03	8.06	114.63	34.4	9.8	4.6	1.5	.8
Stream water, whole	20	6.98	7.79	111.59	33.4	9.0	6.0	1.4	.8
Manure land use									
Spring water, dissolved	2	12.05	7.50	62.20	17.4	17.4	12.0	6.8	6.8
Spring water, whole	2	9.90	9.69	97.85	16.8	16.8	9.9	3.0	3.0
Stream water, dissolved	9	13.03	14.05	107.86	44.1	19.4	6.8	4.6	2.8
Stream water, whole	9	11.94	14.26	119.48	42.4	19.6	5.8	3.0	1.4
Runoff water, dissolved	4	4.80	3.64	75.78	10.0	8.5	3.8	2.2	1.6
Runoff water, whole	4	6.55	9.08	138.64	20.0	16.0	2.6	1.1	1.0
Subsoil	9	361.11	332.39	92.05	1,200.0	425.0	200.0	175.0	150.0
Topsoil	9	461.11	409.86	88.89	1,200.0	725.0	300.0	225.0	150.0
Steer manure	3	3,283.33	246.64	7.51	3,450.0	3,450.0	3,400.0	3,000.0	3,000.0
Cow manure	4	3,625.00	1,088.19	30.02	4,750.0	4,637.5	3,675.0	2,562.5	2,400.0
Pig manure	12	838.45	2,125.53	253.51	7,200.0	142.6	9.8	1.8	1.0
Chicken manure	4	4,812.50	551.32	11.46	5,350.0	5,312.5	4,850.0	4,275.0	4,200.0
Fertilizer land use									
Stream water, dissolved	3	1.55	1.47	94.54	3.2	3.2	1.0	.4	.4
Stream water, whole	3	1.83	1.34	73.20	3.4	3.4	1.4	.8	.8
Runoff water, dissolved	4	1.55	.23	14.90	1.8	1.8	1.6	1.4	1.4
Runoff water, whole	4	1.62	.56	34.58	2.4	2.2	1.6	1.1	1.0
Subsoil	4	387.50	246.22	63.54	750.0	637.5	300.0	225.0	200.0
Topsoil	4	487.50	246.22	50.51	850.0	737.5	400.0	325.0	300.0
Fertilizer	6	53,648.50	40,655.34	75.78	100,050.0	91,687.5	61,275.0	7,660.2	266.0
Sewage land use									
Stream water, dissolved	6	14.03	10.44	74.41	34.8	18.6	10.8	8.1	6.0
Stream water, whole	6	13.50	9.54	70.68	32.4	18.0	10.7	7.0	7.0
Effluent, dissolved	7	18.97	9.84	51.86	37.0	23.4	17.7	9.8	9.0
Effluent, whole	7	18.99	10.45	55.01	39.4	22.8	18.0	9.4	9.4
Sludge	2	6,950.00	2,050.61	29.51	8,400.0	8,400.0	6,950.0	5,500.0	5,500.0
Septic land use									
Stream water, dissolved	6	4.20	2.74	65.17	8.4	6.6	3.9	1.8	1.0
Stream water, whole	6	4.27	3.17	74.33	9.0	7.3	3.6	1.6	1.0
Effluent, dissolved	3	8.18	9.64	117.80	19.2	19.2	4.4	1.0	1.0
Effluent, whole	3	9.73	8.72	89.57	19.8	19.8	4.8	4.6	4.6
Sludge	1	6,700.00	--	--	6,700.0	--	6,700.0	--	6,700.0
Forest land use									
Spring water, dissolved	1	1.00	--	--	1.0	--	1.0	--	1.0
Spring water, whole	1	1.00	--	--	1.0	--	1.0	--	1.0
Stream water, dissolved	7	1.69	1.17	69.62	4.0	2.0	1.0	1.0	.8
Stream water, whole	7	1.63	1.32	81.19	4.4	2.0	1.0	1.0	.4
Subsoil	5	370.00	272.95	73.77	850.0	575.0	300.0	200.0	200.0
Topsoil	6	991.67	796.50	80.32	2,600.0	1,287.5	700.0	575.0	500.0
Precipitation	7	.51	.37	73.65	1.0	1.0	.4	.2	.2

Table 5. Summary statistics for concentrations of carbon, nitrogen, and sulfur in nitrogen sources and associated soil and water in the Lower Susquehanna River Basin, Pennsylvania—Continued

[--, not determined for only one sample; <, less than; ppm, concentrations in parts per million]

Sample class	Number of samples	Mean (ppm)	Standard deviation	Coefficient of variation (percent)	Percentile (ppm)				
					100th Maximum	75th	50th Median	25th	0th Minimum
Organic sulfur (S-org)									
Manure land use									
Subsoil	9	66.67	50.00	75.00	200.0	<100.0	<100.0	<100.0	<100.0
Topsoil	9	100.00	150.00	150.00	500.0	<100.0	<100.0	<100.0	<100.0
Steer manure	3	116.67	76.38	65.47	200.0	200.0	100.0	<100.0	<100.0
Cow manure	4	100.00	70.71	70.71	200.0	175.0	<100.0	<100.0	<100.0
Pig manure	2	750.00	919.24	122.57	1,400.0	1,400.0	--	100.0	100.0
Chicken manure	4	112.50	62.92	55.92	200.0	175.0	100.0	<100.0	<100.0
Fertilizer land use									
Subsoil	5	60.00	22.36	37.27	100.0	<100.0	<100.0	<100.0	<100.0
Topsoil	5	120.00	156.52	130.44	400.0	150.0	<100.0	<100.0	<100.0
Fertilizer	5	7,785.00	16,559.08	212.70	37,400.0	19,150.0	300.0	<100.0	<100.0
Sewage land use									
Sludge	2	850.00	353.55	41.59	1,100.0	1,100.0	850.0	600.0	600.0
Septic land use									
Sludge	1	1,100.0	--	--	1,100.0	--	1,100.0	--	1,100.0
Forest land use									
Subsoil	5	120.00	156.52	130.44	400.0	150.0	<100.0	<100.0	<100.0
Topsoil	6	195.00	117.26	60.13	370.0	317.5	150.0	100.0	100.0
Inorganic sulfur (S-inorg)									
Manure land use									
Subsoil	9	294.44	331.14	112.46	1,150.0	300.0	150.0	125.0	100.0
Topsoil	9	361.11	336.13	93.08	1,150.0	450.0	250.0	175.0	100.0
Steer manure	3	3,166.67	321.46	10.15	3,400.0	3,400.0	3,300.0	2,800.0	2,800.0
Cow manure	4	3,525.00	1,135.41	32.21	4,700.0	4,575.0	3,600.0	2,400.0	2,200.0
Pig manure	2	4,100.00	4,242.64	103.48	7,100.0	7,100.0	4,100.0	1,100.0	1,100.0
Chicken manure	4	4,700.00	605.53	12.88	5,300.0	5,252.0	4,750.0	4,100.0	4,000.0
Fertilizer land use									
Subsoil	4	325.00	221.74	68.23	650.0	550.0	250.0	175.0	150.0
Topsoil	4	350.00	81.65	23.33	450.0	425.0	350.0	275.0	250.0
Fertilizer	5	56,540.00	39,029.05	69.03	99,750.0	94,225.0	59,200.0	17,525.0	9,150.0
Sewage land use									
Sludge	2	6,100.00	2,404.16	39.41	7,800.0	7,800.0	6,100.0	4,400.0	4,400.0
Septic land use									
Sludge	1	5,600.00	--	--	5,600.0	--	5,600.0	--	5,600.0
Forest land use									
Subsoil	5	250.00	122.47	48.99	450.0	350.0	250.0	150.0	150.0
Topsoil	6	796.67	705.31	88.53	2,230.0	1,007.5	525.0	475.0	400.0

Table 5. Summary statistics for concentrations of carbon, nitrogen, and sulfur in nitrogen sources and associated soil and water in the Lower Susquehanna River Basin, Pennsylvania—Continued

[--, not determined for only one sample; <, less than; ppm, concentrations in parts per million]

Sample class	Number of samples	Mean (ppm)	Standard deviation	Coefficient of variation (percent)	Percentile (ppm)				
					100th Maximum	75th	50th Median	25th	0th Minimum
Sulfate sulfur (SO_4-S)									
Mixed land use									
Spring water, dissolved	2	3.70	1.41	38.22	4.7	4.7	3.7	2.7	2.7
Spring water, whole	2	2.35	.49	21.06	2.7	2.7	2.3	2.0	2.0
Stream water, dissolved	20	6.97	8.05	115.43	34.3	9.7	4.5	1.4	<1.0
Stream water, whole	20	6.93	7.79	112.38	33.3	9.0	6.0	1.4	<1.0
Manure land use									
Spring water, dissolved	2	12.00	7.50	62.46	17.3	17.3	12.0	6.7	6.7
Spring water, whole	2	9.85	9.69	98.35	16.7	16.7	9.8	3.0	3.0
Stream water, dissolved	9	12.97	14.04	108.27	44.0	19.3	6.7	4.5	2.7
Stream water, whole	9	11.88	14.25	119.99	42.3	19.5	5.7	3.0	1.3
Runoff water, dissolved	4	4.75	3.64	76.57	10.0	8.4	3.7	2.1	1.6
Runoff water, whole	4	6.50	9.08	139.70	20.0	15.9	2.5	1.1	1.0
Subsoil	9	244.44	331.14	135.47	1,100.0	250.0	100.0	<100.0	<100.0
Topsoil	9	311.11	336.13	108.04	1,100.0	400.0	200.0	<100.0	<100.0
Steer manure	3	1,000.00	1,307.67	130.77	2,500.0	2,500.0	400.0	100.0	100.0
Cow manure	4	1,225.00	1,472.81	120.23	3,200.0	2,775.0	800.0	100.0	100.0
Pig manure	12	121.57	216.39	177.99	700.0	142.5	9.0	1.7	<1.0
Chicken manure	4	1,600.00	1,465.15	91.57	3,500.0	3,100.0	1,400.0	300.0	100.0
Fertilizer land use									
Stream water, dissolved	3	1.43	1.40	97.76	3.0	3.0	1.0	<1.0	<1.0
Stream water, whole	3	1.77	1.36	77.06	3.3	3.3	1.3	<1.0	<1.0
Runoff water, dissolved	4	1.50	.23	15.40	1.7	1.7	1.5	1.3	1.3
Runoff water, whole	4	1.57	.56	35.68	2.3	2.1	1.5	1.1	1.0
Subsoil	5	240.00	207.36	86.40	600.0	400.0	200.0	100.0	100.0
Topsoil	5	300.00	70.71	23.57	400.0	350.0	300.0	250.0	200.0
Fertilizer	6	38,625.00	40,075.84	103.76	99,700.0	75,550.0	29,225.0	4,442.5	265.0
Sewage land use									
Stream water, dissolved	6	13.98	10.44	74.67	34.7	18.5	10.7	8.0	6.0
Stream water, whole	6	13.45	9.54	70.94	32.3	17.9	10.6	7.0	7.0
Effluent, dissolved	7	18.91	9.83	52.00	37.0	23.3	17.7	9.7	9.0
Effluent, whole	7	18.94	10.45	55.15	39.3	22.7	18.0	9.3	9.3
Sludge	2	600.00	424.26	70.71	900.0	900.0	600.0	300.0	300.0
Septic land use									
Stream water, dissolved	6	4.15	2.74	65.95	8.3	6.6	3.8	1.0	<1.0
Stream water, whole	6	4.22	3.17	75.21	9.0	7.3	3.5	<1.0	<1.0
Effluent, dissolved	3	8.13	9.64	118.52	19.1	19.1	4.3	1.0	1.0
Effluent, whole	3	9.13	7.77	85.02	18.1	18.1	4.7	4.6	4.6
Sludge	1	2,300.00	--	--	2,300.0	--	2,300.0	--	2,300.0
Forest land use									
Spring water, dissolved	1	<1.00	--	--	<1.0	--	<1.0	--	<1.0
Spring water, whole	1	<1.00	--	--	<1.0	--	<1.0	--	<1.0
Stream water, dissolved	7	1.63	1.18	72.46	4.0	2.0	<1.0	<1.0	<1.0
Stream water, whole	7	1.57	1.33	84.64	4.3	2.0	<1.0	<1.0	<1.0
Subsoil	5	200.00	122.47	61.24	400.0	300.0	200.0	100.0	100.0
Topsoil	6	293.33	121.11	41.29	500.0	395.0	250.0	200.0	200.0
Precipitation	7	.46	.37	81.71	1.0	1.0	<1.0	<1.0	<1.0

Table 5. Summary statistics for concentrations of carbon, nitrogen, and sulfur in nitrogen sources and associated soil and water in the Lower Susquehanna River Basin, Pennsylvania—Continued

[--, not determined for only one sample; <, less than; ppm, concentrations in parts per million]

Sample class	Number of samples	Mean (ppm)	Standard deviation	Coefficient of variation (percent)	Percentile (ppm)				
					100th Maximum	75th	50th Median	25th	0th Minimum
				Sulfide sulfur (H_2S-S)					
Mixed land use									
Spring water, dissolved	2	<0.10	--	--	<0.1	<0.1	<0.1	<0.1	<0.1
Spring water, whole	2	<.10	--	--	<.1	<.1	<.1	<.1	<.1
Stream water, dissolved	20	<.10	.02	27.98	.1	<.1	<.1	<.1	<.1
Stream water, whole	20	<.10	.02	27.98	.1	<.1	<.1	<.1	<.1
Manure land use									
Spring water, dissolved	2	<.10	--	--	<.1	<.1	<.1	<.1	<.1
Spring water, whole	2	<.10	--	--	<.1	<.1	<.1	<.1	<.1
Stream water, dissolved	9	.06	.02	36.08	.1	<.1	<.1	<.1	<.1
Stream water, whole	9	.06	.02	36.08	.1	<.1	<.1	<.1	<.1
Runoff water, dissolved	4	<.10	--	--	<.1	<.1	<.1	<.1	<.1
Runoff water, whole	4	<.10	--	--	<.1	<.1	<.1	<.1	<.1
Subsoil	9	<100.00	--	--	<100.0	<100.0	<100.0	<100.0	<100.0
Topsoil	9	<100.00	--	--	<100.0	<100.0	<100.0	<100.0	<100.0
Steer manure	3	2,166.67	1,628.91	75.18	3,300.0	3,300.0	2,900.0	300.0	300.0
Cow manure	4	2,300.00	1,232.88	53.60	4,100.0	3,600.0	1,800.0	1,500.0	1,500.0
Pig manure	12	591.87	1,926.99	325.57	6,700.0	.8	.1	<.1	<.1
Chicken manure	4	3,100.00	1,122.50	36.21	4,300.0	4,075.0	3,250.0	1,975.0	1,600.0
Fertilizer land use									
Stream water, dissolved	3	.12	.08	65.47	.2	.2	.1	<.1	<.1
Stream water, whole	3	.07	.03	43.30	.1	.1	<.1	<.1	<.1
Runoff water, dissolved	4	<.10	--	--	<.1	<.1	<.1	<.1	<.1
Runoff water, whole	4	<.10	--	--	<.1	<.1	<.1	<.1	<.1
Subsoil	4	<100.00	--	--	<100.0	<100.0	<100.0	<100.0	<100.0
Topsoil	4	<100.00	--	--	<100.0	<100.0	<100.0	<100.0	<100.0
Fertilizer	6	8,536.00	8,856.01	103.75	21,200.0	17,262.5	7,007.5	<100.0	<100.0
Sewage land use									
Stream water, dissolved	6	<.10	--	--	<.1	<.1	<.1	<.1	<.1
Stream water, whole	6	<.10	--	--	<.1	<.1	<.1	<.1	<.1
Effluent, dissolved	7	.06	.02	33.07	.1	<.1	<.1	<.1	<.1
Effluent, whole	7	<.10	--	--	<.1	<.1	<.1	<.1	<.1
Sludge	2	5,500.00	1,979.90	36.00	6,900.0	6,900.0	5,500.0	4,100.0	4,100.0
Septic land use									
Stream water, dissolved	6	<.10	--	--	<.1	<.1	<.1	<.1	<.1
Stream water, whole	6	<.10	--	--	<.1	<.1	<.1	<.1	<.1
Effluent, dissolved	3	<.10	--	--	<.1	<.1	<.1	<.1	<.1
Effluent, whole	3	.60	.95	158.77	1.7	1.7	<.1	<.1	<.1
Sludge	1	3,300.00	--	--	3,300.0	--	3,300.0	--	3,300.0
Forest land use									
Spring water, dissolved	1	<.10	--	--	<.1	--	<.1	--	<.1
Spring water, whole	1	<.10	--	--	<.1	--	<.1	--	<.1
Stream water, dissolved	7	.06	.02	33.07	.1	<.1	<.1	<.1	<.1
Stream water, whole	7	.06	.02	33.07	.1	<.1	<.1	<.1	<.1
Subsoil	5	<100.00	--	--	<100.0	<100.0	<100.0	<100.0	<100.0
Topsoil	6	503.33	676.66	134.44	1,870.0	692.5	300.0	125.0	<100.0
Precipitation	7	<.10	--	--	<.1	<.1	<.1	<.1	<.1

Table 5. Summary statistics for concentrations of carbon, nitrogen, and sulfur in nitrogen sources and associated soil and water in the Lower Susquehanna River Basin, Pennsylvania—Continued

[--, not determined for only one sample; <, less than; ppm, concentrations in parts per million]

Sample class	Number of samples	Mean (ppm)	Standard deviation	Coefficient of variation (percent)	Percentile (ppm) 100th Maximum	75th	50th Median	25th	0th Minimum
Molar ratio of total carbon to total nitrogen (C:N)									
Mixed land use									
Spring water, whole	2	23.22	3.76	16.19	25.9	25.9	23.2	20.6	20.6
Stream water, whole	20	12.72	8.20	64.48	31.1	16.4	11.9	6.8	.9
Manure land use									
Spring water, whole	2	11.29	12.47	110.49	20.1	20.1	11.3	2.5	2.5
Stream water, whole	8	8.84	4.85	54.95	19.5	10.5	7.4	6.3	3.5
Runoff water, whole	4	17.47	12.09	69.24	31.6	29.5	16.6	6.3	5.1
Subsoil	9	10.64	1.92	18.00	12.8	12.5	10.4	8.8	7.7
Topsoil	9	12.45	4.12	33.07	22.8	12.7	12.0	9.9	8.9
Steer manure	3	13.77	.63	4.60	14.3	14.3	13.9	13.1	13.1
Cow manure	4	16.05	2.90	18.04	20.0	19.1	15.2	13.8	13.7
Pig manure	7	5.51	5.75	104.48	15.4	11.4	3.0	.8	.3
Chicken manure	4	7.29	2.39	32.78	9.8	9.5	7.3	5.0	4.8
Fertilizer land use									
Stream water, whole	3	5.30	2.08	39.26	7.4	7.4	5.2	3.2	3.2
Runoff water, whole	4	13.83	13.01	94.07	25.9	25.5	13.7	2.3	2.0
Subsoil	5	11.27	1.51	13.38	13.8	12.6	10.7	10.2	10.0
Topsoil	5	12.02	3.36	27.99	14.7	14.6	12.4	9.3	6.4
Fertilizer	6	.53	.50	95.75	1.3	1.1	.3	.1	<.1
Sewage land use									
Stream water, whole	6	8.57	3.72	43.40	14.2	10.9	9.0	5.4	3.3
Effluent, whole	7	4.50	2.86	63.57	8.2	7.5	3.1	1.6	1.6
Sludge	2	8.78	1.48	16.80	9.8	9.8	8.8	7.7	7.7
Septic land use									
Stream water, whole	6	14.65	8.02	54.75	28.8	21.8	11.3	8.5	8.5
Effluent, whole	3	5.75	5.54	96.38	12.2	12.2	2.6	2.5	2.5
Sludge	1	7.59	--	--	7.6	--	7.6	--	7.6
Forest land use									
Spring water, whole	1	21.85	--	--	21.8	--	21.8	--	21.8
Stream water, whole	7	7.69	3.41	44.40	12.1	10.1	8.2	6.5	1.3
Subsoil	5	24.35	4.99	20.51	29.5	29.3	23.7	19.7	17.8
Topsoil	6	29.11	2.24	7.69	32.8	31.2	28.4	27.2	27.1

Table 5. Summary statistics for concentrations of carbon, nitrogen, and sulfur in nitrogen sources and associated soil and water in the Lower Susquehanna River Basin, Pennsylvania—Continued

[--, not determined for only one sample; <, less than; ppm, concentrations in parts per million]

Sample class	Number of samples	Mean (ppm)	Standard deviation	Coefficient of variation (percent)	Percentile (ppm)				
					100th Maximum	75th	50th Median	25th	0th Minimum
Molar ratio of organic carbon to total nitrogen (C-org:N)									
Mixed land use									
Spring water, whole	2	2	5.24	1.91	36.50	6.6	6.6	5.2	3.9
Stream water, whole	20	20	2.78	1.94	69.95	6.5	4.6	2.4	1.1
Manure land use									
Spring water, whole	2	2	.17	.16	92.46	.3	.3	.2	<.1
Stream water, whole	8	8	4.05	5.97	147.26	18.4	4.0	2.3	.6
Runoff water, whole	4	4	5.88	5.38	91.52	13.9	11.4	3.8	2.5
Subsoil	9	9	10.29	1.92	18.65	12.5	12.0	10.4	8.3
Topsoil	9	9	12.02	4.11	34.17	22.4	12.1	11.7	9.6
Steer manure	3	3	13.76	.63	4.61	14.3	14.3	13.9	13.1
Cow manure	4	4	15.23	2.16	14.19	18.1	17.5	14.6	13.6
Pig manure	7	7	5.22	5.92	113.46	15.4	11.3	2.3	.8
Chicken manure	4	4	7.26	2.38	32.71	9.8	9.5	7.3	5.0
Fertilizer land use									
Stream water, whole	3	3	2.51	1.51	59.92	4.2	4.2	2.1	1.3
Runoff water, whole	4	4	8.12	8.29	102.14	17.3	16.2	7.3	.9
Subsoil	5	5	10.96	1.65	15.06	13.7	12.3	10.6	9.7
Topsoil	5	5	11.28	3.44	30.51	14.5	14.4	11.0	8.2
Fertilizer	6	6	.47	.43	92.85	1.2	.9	.3	.1
Sewage land use									
Stream water, whole	6	6	2.82	2.48	88.05	7.3	4.4	2.3	.8
Effluent, whole	7	7	.86	.45	51.78	1.6	1.4	.6	.5
Sludge	2	2	8.69	1.41	16.20	9.7	9.7	8.7	7.7
Septic land use									
Stream water, whole	6	6	3.61	2.14	59.26	5.8	5.6	3.7	1.8
Effluent, whole	3	3	4.50	5.03	111.82	10.2	9.6	2.4	1.2
Sludge	1	1	7.51	--	--	7.5	--	7.5	--
Forest land use									
Spring water, whole	1	1	1.06	--	--	1.1	--	1.1	--
Stream water, whole	7	7	5.09	3.00	59.06	9.2	8.2	5.1	2.0
Subsoil	5	5	24.06	5.04	20.93	29.5	29.1	23.4	19.3
Topsoil	6	6	28.89	2.27	7.84	32.7	31.1	28.0	27.1

Table 6. Summary statistics for $\delta^{13}C$, $\delta^{15}N$, and $\delta^{34}S$ data for nitrogen sources and associated soil and water in the Lower Susquehanna River Basin, Pennsylvania

[--, not determined]

Sample class	Number of samples	Mean (per mil)	Standard deviation	Percentile (per mil)				
				100th Maximum	75th	50th Median	25th	0th Minimum
$\delta^{13}C$ in total carbon								
Mixed land use								
Spring, particulate	2	-27.00	2.90	-25.0	-25.0	-27.0	-29.0	-29.0
Stream, particulate	20	-26.04	2.30	-20.7	-24.9	-25.8	-27.5	-31.2
Manure land use								
Spring, particulate	2	-24.15	3.46	-21.7	-21.7	-24.2	-26.6	-26.6
Stream, particulate	8	-22.81	1.80	-20.3	-21.0	-23.1	-24.1	-25.6
Runoff particulate	4	-21.05	3.53	-17.9	-18.4	-20.1	-24.7	-26.1
Subsoil	9	-19.76	1.34	-18.0	-18.5	-19.7	-21.2	-21.4
Topsoil	9	-19.21	1.76	-16.1	-18.3	-19.1	-20.7	-22.1
Steer manure	3	-13.47	1.10	-12.2	-12.2	-14.0	-14.2	-14.2
Cow manure	4	-19.49	1.04	-18.3	-18.5	-19.5	-20.5	-20.6
Pig manure	6	-16.42	1.66	-14.9	-15.0	-16.0	-18.3	-18.4
Chicken manure	4	-18.20	.42	-17.8	-17.8	-18.1	-18.6	-18.7
Fertilizer land use								
Stream, particulate	3	-25.33	.93	-24.3	-24.3	-25.6	-26.1	-26.1
Runoff, particulate	4	-24.66	4.18	-18.6	-20.3	-26.2	-27.6	-27.7
Subsoil	5	-21.57	2.51	-19.5	-19.6	-20.1	-24.3	-24.5
Topsoil	5	-21.65	2.71	-19.0	-19.2	-20.7	-24.5	-24.8
Fertilizer	5	-30.44	10.56	-20.6	-22.2	-24.1	-41.9	-42.5
Sewage land use								
Stream, particulate	6	-24.63	1.09	-23.2	-23.4	-25.0	-25.6	-25.7
Effluent, particulate	9	-22.19	1.09	-20.8	-21.4	-22.2	-22.9	-24.3
Sludge	4	-21.89	.52	-21.4	-21.5	-21.8	-22.4	-22.6
Septic land use								
Stream, particulate	6	-26.98	.37	-26.5	-26.6	-27.0	-27.3	-27.5
Effluent, particulate	3	-22.67	.75	-21.9	-21.9	-22.7	-23.4	-23.4
Sludge	1	-24.10	--	-24.1	--	-24.1	--	-24.1
Forest land use								
Spring, particulate	1	-30.30	--	-30.3	--	-30.3	--	-30.3
Stream, particulate	7	-27.31	.39	-26.8	-26.9	-27.3	-27.7	-27.7
Subsoil	5	-26.52	.38	-26.1	-26.2	-26.5	-26.9	-26.9
Topsoil	6	-27.15	.51	-26.4	-26.6	-27.3	-27.5	-27.8

Table 6. Summary statistics for $\delta^{13}C$, $\delta^{15}N$, and $\delta^{34}S$ data for nitrogen sources and associated soil and water in the Lower Susquehanna River Basin, Pennsylvania—Continued

[--, not determined]

Sample class	Number of samples	Mean (per mil)	Standard deviation	Percentile (per mil)				
				100th Maximum	75th	50th Median	25th	0th Minimum
$\delta^{15}N$ in total inorganic nitrogen (mass-weighted average of $\delta^{15}N$-NO_3 and $\delta^{15}N$-NH_3)								
Mixed land use								
Spring, particulate	2	4.20	0.99	4.9	4.9	4.2	3.5	3.5
Stream, dissolved	16	5.93	1.34	8.7	6.9	6.0	5.0	3.4
Stream, particulate	20	4.17	1.98	8.6	5.1	4.3	2.6	.7
Manure land use								
Spring, dissolved	2	9.30	1.67	10.5	10.5	9.3	8.1	8.1
Spring, particulate	2	5.65	.64	6.1	6.1	5.6	5.2	5.2
Stream, dissolved	4	14.50	11.16	29.6	26.2	11.7	5.5	5.0
Stream, particulate	8	6.07	1.92	10.2	7.2	5.2	4.8	4.7
Runoff, dissolved	5	13.34	16.55	41.6	27.3	8.5	1.8	-.1
Runoff, particulate	5	8.41	2.02	11.8	10.0	7.8	7.1	6.4
Subsoil	9	7.29	2.81	11.2	9.4	7.7	5.4	1.8
Topsoil	9	8.59	2.88	12.3	10.9	8.7	6.1	4.0
Steer manure	3	10.80	1.14	11.6	11.6	11.3	9.5	9.5
Cow manure	4	7.41	2.44	9.1	9.0	8.4	4.9	3.8
Pig manure	9	17.22	11.35	37.2	27.9	12.5	8.7	6.3
Chicken manure	4	4.01	4.22	10.2	8.4	2.4	1.2	1.1
Fertilizer land use								
Stream, dissolved	3	7.36	1.72	9.2	9.2	7.1	5.8	5.8
Stream, particulate	3	4.43	2.36	6.9	6.9	4.2	2.2	2.2
Runoff, dissolved	2	2.60	2.96	4.7	4.7	2.6	.5	.5
Runoff, particulate	4	3.87	1.98	5.4	5.3	4.5	1.8	1.1
Subsoil	5	4.14	.61	4.6	4.6	4.5	3.5	3.4
Topsoil	5	2.77	1.03	4.0	3.8	2.5	1.9	1.5
Fertilizer	6	.01	1.80	3.1	1.3	-.3	-1.2	-2.2
Sewage land use								
Stream, dissolved	4	10.46	3.94	15.5	14.4	10.2	6.8	6.0
Stream, particulate	6	7.49	3.10	13.0	10.1	6.1	5.3	5.0
Effluent, dissolved	4	20.37	5.38	26.5	25.4	20.6	15.1	13.7
Effluent, particulate	9	4.63	4.84	10.2	8.2	7.4	.3	-3.7
Sludge	4	6.95	4.63	11.1	10.8	7.8	2.1	1.0
Septic land use								
Stream, dissolved	4	4.30	.42	4.9	4.7	4.2	4.0	3.9
Stream, particulate	6	2.54	1.22	3.7	3.5	2.9	1.5	.4
Effluent, dissolved	2	16.29	21.06	31.2	31.2	16.3	1.4	1.4
Effluent, particulate	3	1.13	1.88	3.0	3.0	1.1	-.7	-.7
Sludge	1	-2.10	--	-2.1	--	-2.1	--	-2.1
Forest land use								
Spring, particulate	1	-1.60	--	-1.6	--	-1.6	--	-1.6
Stream, dissolved	3	3.12	1.07	4.3	4.3	2.9	2.2	2.2
Stream, particulate	7	.70	.93	1.7	1.3	.9	.2	-1.1
Subsoil	5	4.70	.54	5.3	5.2	4.6	4.2	3.9
Topsoil	6	.27	.36	.6	.6	.3	-.1	-.3

Table 6. Summary statistics for $\delta^{13}C$, $\delta^{15}N$, and $\delta^{34}S$ data for nitrogen sources and associated soil and water in the Lower Susquehanna River Basin, Pennsylvania—Continued

[--, not determined]

Sample class	Number of samples	Mean (per mil)	Standard deviation	Percentile (per mil)				
				100th Maximum	75th	50th Median	25th	0th Minimum
$\delta^{15}N$-NO_3 in total inorganic nitrogen								
Mixed land use								
Stream, dissolved	16	5.82	1.44	8.7	6.8	5.8	5.0	2.6
Manure land use								
Spring, dissolved	2	9.45	1.77	10.7	10.7	9.4	8.2	8.2
Stream, dissolved	5	14.61	14.54	39.3	27.7	7.3	5.1	5.0
Runoff, dissolved	4	3.20	4.04	8.5	7.4	2.2	-.1	-.2
Pig manure	2	11.70	.99	12.4	12.4	11.7	11.0	11.0
Fertilizer land use								
Stream, dissolved	3	7.63	1.72	9.2	9.2	7.9	5.8	5.8
Runoff, dissolved	2	2.10	3.68	4.7	4.7	2.1	-.5	-.5
Fertilizer	1	-2.00	--	-2.0	--	-2.0	--	-2.0
Sewage land use								
Stream, dissolved	4	9.67	4.24	15.7	14.1	8.5	6.4	6.0
Effluent, dissolved	6	14.57	9.32	26.7	24.6	12.7	7.7	1.6
Septic land use								
Stream, dissolved	4	4.30	.42	4.9	4.7	4.2	4.0	3.9
Effluent, dissolved	2	1.30	2.83	3.3	3.3	1.3	-.7	-.7
Forest land use								
Stream, dissolved	3	3.43	1.10	4.3	4.3	3.8	2.2	2.2
$\delta^{15}N$-NH_3 in total inorganic nitrogen								
Mixed land use								
Stream, dissolved	3	5.67	1.48	7.3	7.3	5.3	4.4	4.4
Manure land use								
Spring, dissolved	2	-.40	2.97	1.7	1.7	-.4	-2.5	-2.5
Stream, dissolved	4	17.52	10.79	31.1	27.6	17.1	7.8	4.7
Runoff, dissolved	4	14.67	19.06	42.1	34.8	8.2	1.0	.2
Pig manure	2	35.50	2.40	37.2	37.2	35.5	33.8	33.8
Fertilizer land use								
Stream, dissolved	1	3.00	--	3.0	--	3.0	--	3.0
Runoff, dissolved	1	8.60	--	8.6	--	8.6	--	8.6
Fertilizer	1	-2.40	--	-2.4	--	-2.4	--	-2.4
Sewage land use								
Stream, dissolved	2	10.15	8.13	15.9	15.9	10.1	4.4	4.4
Effluent, dissolved	5	16.20	7.95	22.9	21.1	18.4	10.2	2.4
Septic land use								
Effluent, dissolved	2	16.30	21.07	31.2	31.2	16.3	1.4	1.4
Forest land use								
Stream, dissolved	1	1.30	--	1.3	--	1.3	--	1.3

Table 6. Summary statistics for $\delta^{13}C$, $\delta^{15}N$, and $\delta^{34}S$ data for nitrogen sources and associated soil and water in the Lower Susquehanna River Basin, Pennsylvania—Continued

[--, not determined]

Sample class	Number of samples	Mean (per mil)	Standard deviation	Percentile (per mil)				
				100th Maximum	75th	50th Median	25th	0th Minimum
			$\delta^{34}S$ in total sulfur					
Mixed land use								
Spring, dissolved	2	5.55	1.06	6.3	6.3	5.5	4.8	4.8
Spring, particulate	1	4.80	--	4.8	--	4.8	--	4.8
Stream, dissolved	20	6.97	3.06	12.7	9.8	5.5	4.4	2.7
Stream, particulate	17	5.17	2.59	13.0	5.3	4.7	3.4	2.5
Manure land use								
Spring, dissolved	2	3.70	0	3.7	3.7	3.7	3.7	3.7
Spring, particulate	2	4.65	.92	5.3	5.3	4.6	4.0	4.0
Stream, dissolved	8	5.69	2.42	11.2	6.3	4.9	4.6	3.2
Stream, particulate	7	4.99	1.33	7.7	5.3	4.7	4.2	3.6
Runoff, dissolved	3	2.33	1.88	4.0	4.0	2.7	.3	.3
Runoff, particulate	5	3.62	.74	4.4	4.1	3.7	3.0	2.4
Subsoil	9	4.96	1.81	7.6	6.7	4.1	3.4	2.6
Topsoil	9	3.72	3.59	6.5	5.9	4.5	3.5	-5.4
Steer manure	3	6.00	.72	6.8	6.8	5.8	5.4	5.4
Cow manure	4	4.21	.67	5.0	4.8	4.2	3.6	3.4
Pig manure	10	3.69	1.77	5.8	4.5	3.9	3.5	-.9
Chicken manure	4	3.35	.71	3.8	3.8	3.6	2.6	2.3
Fertilizer land use								
Stream, dissolved	3	4.22	.63	4.8	4.8	4.2	3.6	3.6
Stream, particulate	2	4.10	.57	4.5	4.5	4.1	3.7	3.7
Runoff, dissolved	3	5.20	2.77	8.4	8.4	3.6	3.6	3.6
Runoff, particulate	4	5.84	1.61	8.2	7.5	5.2	4.8	4.8
Subsoil	5	5.66	.83	6.6	6.5	5.3	4.9	4.8
Topsoil	5	4.83	1.51	6.8	6.4	4.0	3.6	3.3
Fertilizer	6	8.22	6.61	19.9	11.2	7.9	4.2	-.5
Sewage land use								
Stream, dissolved	6	5.42	2.49	8.0	7.2	5.8	4.1	.8
Stream, particulate	6	3.76	1.17	5.1	4.8	3.8	2.5	2.5
Effluent, dissolved	8	4.97	1.11	6.4	5.8	5.1	3.7	3.5
Effluent, particulate	9	4.78	.78	6.1	5.2	4.8	4.2	3.6
Sludge	4	2.47	.65	3.4	3.1	2.3	2.0	1.9
Septic land use								
Stream, dissolved	5	5.40	1.10	7.0	6.5	4.7	4.6	4.5
Stream, particulate	5	4.56	1.45	6.1	5.8	4.5	3.3	2.3
Effluent, dissolved	2	9.20	.28	9.4	9.4	9.2	9.0	9.0
Effluent, particulate	3	3.91	.35	4.1	4.1	4.1	3.5	3.5
Sludge	1	1.00	--	1.0	--	1.0	--	1.0
Forest land use								
Stream, dissolved	7	3.94	1.07	4.9	4.6	4.4	3.6	1.7
Stream, particulate	7	4.60	1.68	7.8	5.1	4.5	3.1	2.6
Subsoil	5	6.40	2.11	9.8	8.1	6.0	4.9	4.0
Topsoil	6	3.57	.98	4.6	4.5	3.7	2.7	2.0
Precipitation	2	1.70	3.96	4.5	4.5	1.7	-1.1	-1.1

however, and after boiling acidified samples to concentrate the solutes, only S isotopic compositions could be measured in two samples with $\delta^{34}S$ of –1.1 and +4.5‰ (table 6, fig. 7). Other researchers found similar results for S (Nriagu and Coker, 1978; Stam and others, 1992). Previous workers reported rainfall NH_4^+ and NO_3^- has a wide range of $\delta^{15}N$ from –13.7 to +9.0‰ and no clear trend of ^{15}N enrichment in either ion (Hoering, 1957; Moore, 1974; Freyer, 1978).

Forest Leaf Litter

Topsoil samples collected in forested areas in Stony Creek and Dogwood Run Basins (fig. 1) consist of dark-brown, humus-rich organic matter that contains total C of 15.5 to 49.8 wt %, total N of 0.6 to 2.1 wt %, and molar C-org:N of 27 to 33 (table 5). Organic C composes nearly 100 percent of the total C; organic N composes 81 to 86 percent of total N, and inorganic NH_3-N and NO_3-N composes the balance (fig. 4). The isotopic composition of forest leaf litter is generally distinctive and less variable than other N sources. Forest leaf litter collected for this study (table 6, fig. 7) has $\delta^{13}C$ of –27.8 to –26.4‰, $\delta^{15}N$ of –0.3 to +0.6‰, and $\delta^{34}S$ of 2.0 to 4.7‰. Leaves of upland trees, which form litter input to the forest soil, characteristically have $\delta^{13}C$ of –30‰ to –22‰ (Smith and Epstein, 1971; Deines, 1980) and $\delta^{15}N$ of –8 to +3‰ (Peterson and Fry, 1987; Nadelhoffer and Fry, 1988). The C and N isotopic compositions of forest leaf litter collected for this study are consistent with those expected of the source leaves. The litter S isotopic composition is consistent with that of precipitation.

Synthetic Fertilizer

Synthetic fertilizers typically are produced by reacting H_2 gas with atmospheric N_2 at high temperature and pressure (Rochow, 1977; Teply and others, 1980). The resulting NH_3 is then reacted with various acids or CO_2 to produce ammonium salts (Teply and others, 1980). The liquid fertilizer sample (BE1FL in tables B3 and B4) collected at the Bald Eagle Creek farm site is "30-percent-N." Typical liquid fertilizers containing 28 to 32 percent N are produced commercially by mixing a 75-percent solution of ammonium carbamate (NH_2COONH_4) and an 83-percent solution of ammonium-nitrate (NH_4NO_3) (Teply and others, 1980, p. 21). The four solid fertilizer samples (BE1FS, M2AFS, M2AFS1, M2AFS2 in tables B3 and B4) collected at the Bald Eagle farm and the Monroe Creek golf-course sites have different concentrations, expressed as percent, of total N (N), available phosphoric acid (P_2O_5), and soluble potash (K_2O). Sample BE1FS is a 15-15-15 (N-P_2O_5-K_2O) mixture of solid urea (NH_2CONH_2), ammonium sulfate [$(NH_4)_2SO_4$], diammonium phosphate [$(NH_4)_2HPO_4$], plus potassium salts such as potassium chloride (KCl). Samples M2AFS, M2AFS1, and M2AFS2 are 34-6-7, 34-3-8, and 18-4-10 mixtures, respectively, of urea plus smaller quantities of NH_4NO_3 or $(NH_4)_2SO_4$ plus $(NH_4)_2HPO_4$.

The highest concentrations of N in the different source materials collected for this study were found in the synthetic fertilizer samples (tables 5 and B3). The predominant forms of C, N, and S in the fertilizer samples are organic-C, organic-N or NH_3, and SO_4 (fig. 4). The concentrations of C and S differ in each sample because of the various compositions of the synthetic N compounds. Synthetic urea and ammonium carbamate contain organic C, and ammonium sulfate contains S as SO_4. Because the ultimate source of N in the synthetic fertilizers is atmospheric N_2 with $\delta^{15}N$ of 0‰, values of $\delta^{15}N$ for the synthetic fertilizers also are expected to be about 0‰. The liquid fertilizer, BE1FL, has $\delta^{15}N$-NO_3 of –2.0‰ and $\delta^{15}N$-NH_3 of –2.4‰, and $\delta^{34}S$ of 5.8‰. The solid fertilizers have similar $\delta^{15}N$ of –0.5 to 4.3‰, but variable $\delta^{13}C$ of –42.5 to –20.6‰ and $\delta^{34}S$ of –0.5 to +19.9‰ (tables 5 and B4). These ranges include extreme values for the lowest $\delta^{13}C$ and highest $\delta^{34}S$ measured in this study. A probable source of isotopically light C is petroleum, which is used in fertilizer synthesis. Probable sources of isotopically light or heavy S include elemental and petroleum S, both of which are used in the manufacture of sulfuric acid to produce ammonium sulfate (Rochow, 1977).

Manure

Four general varieties of farm-animal manures were collected: chicken, swine (pig), dairy cattle (cow), and feeder cattle (steer). These animals represent the majority of livestock raised in the Lower Susquehanna River Basin (Ott and others, 1991); most of the remainder are turkeys, horses, sheep, and goats. Fresh animal manures consist mostly of water (60 to 85 wt %) and partially degraded organic compounds including remnants of the feed and microorganism

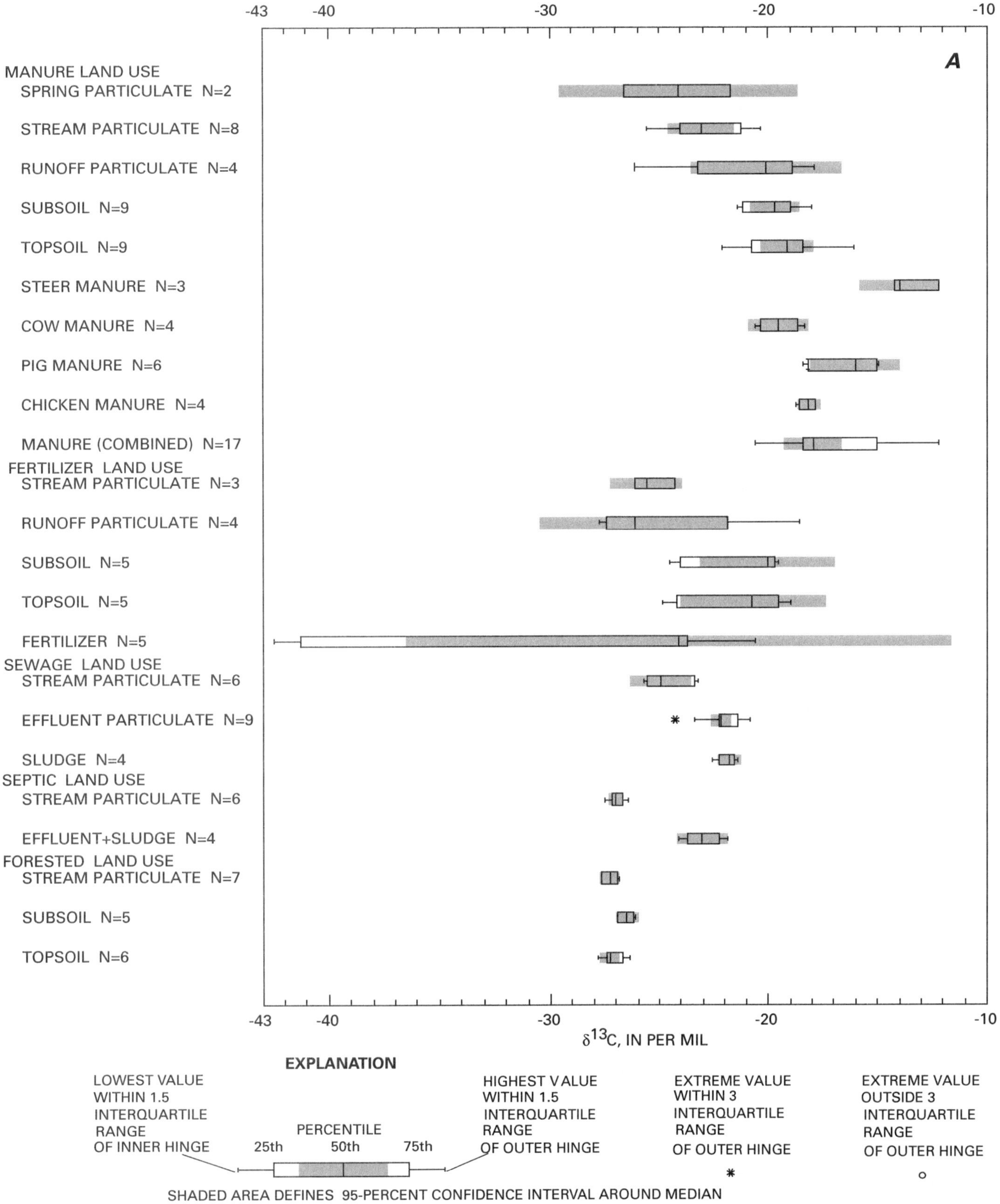

Figure 7. $\delta^{13}C$, $\delta^{15}N$, $\delta^{34}S$, and molar ratio of organic-C to total-N (C-org:N) in N-source, soil, and water samples, Lower Susquehanna River Basin, Pennsylvania.

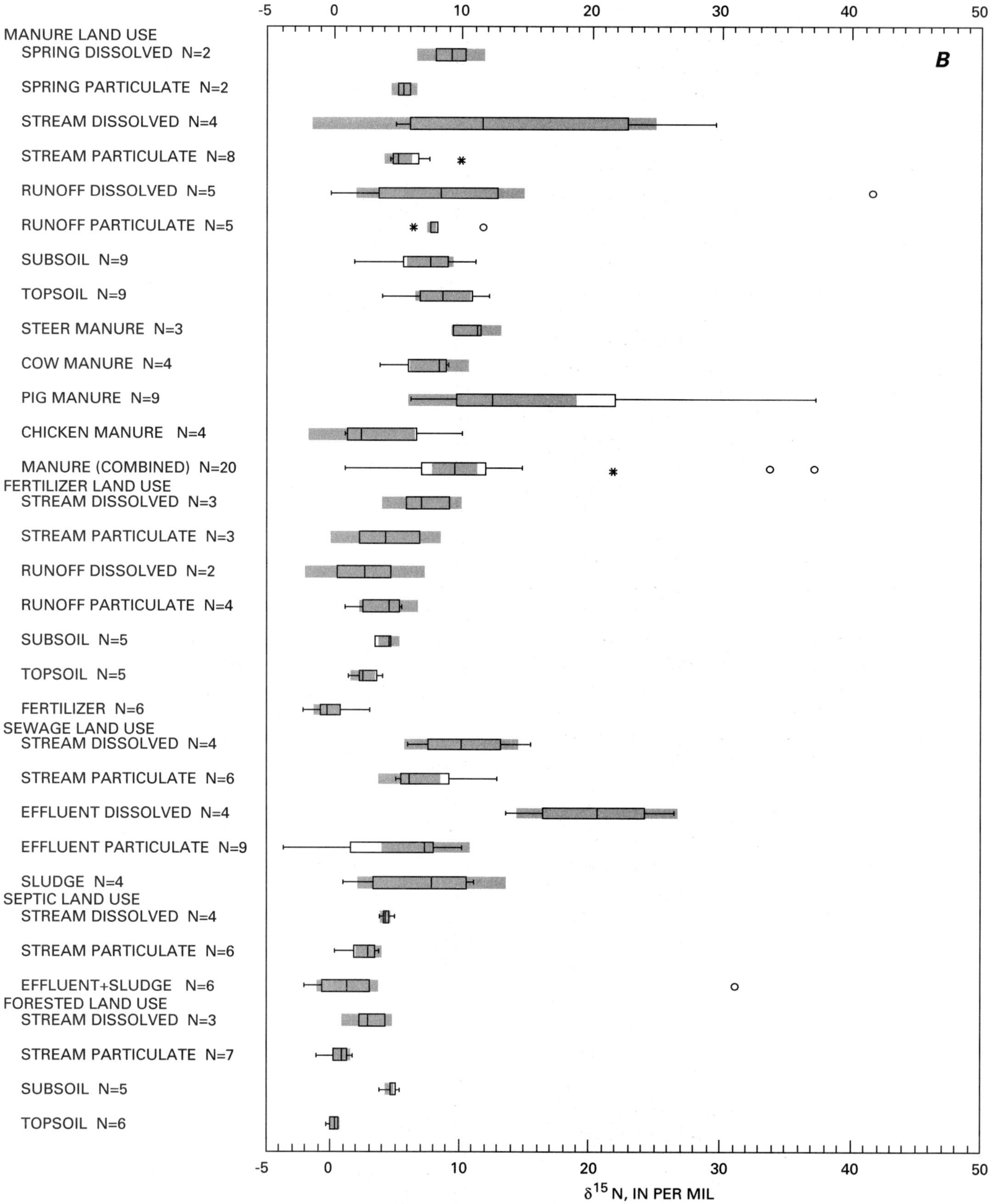

Figure 7. $\delta^{13}C$, $\delta^{15}N$, $\delta^{34}S$, and molar ratio of organic-C to total-N (C-org:N) in N-source, soil, and water samples, Lower Susquehanna River Basin, Pennsylvania—Continued.

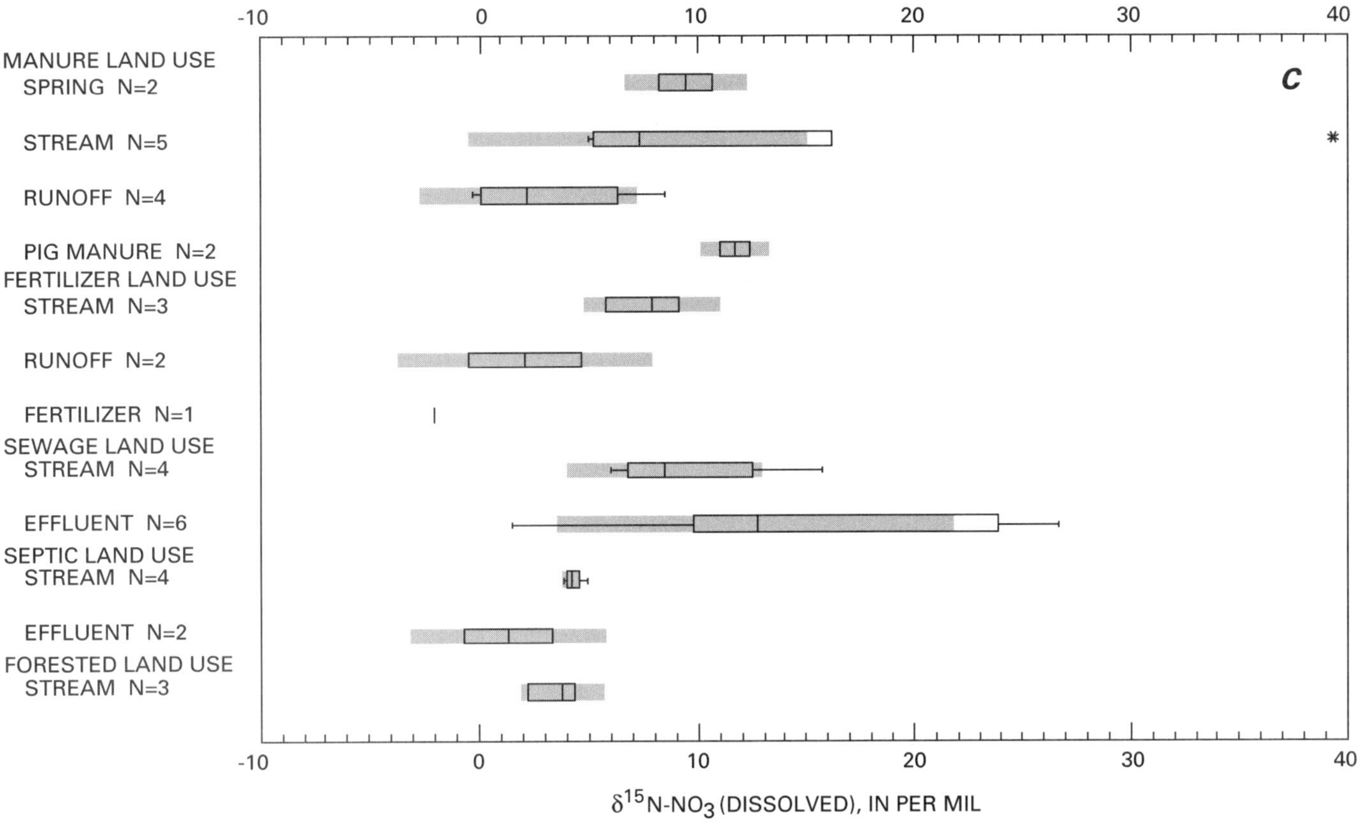

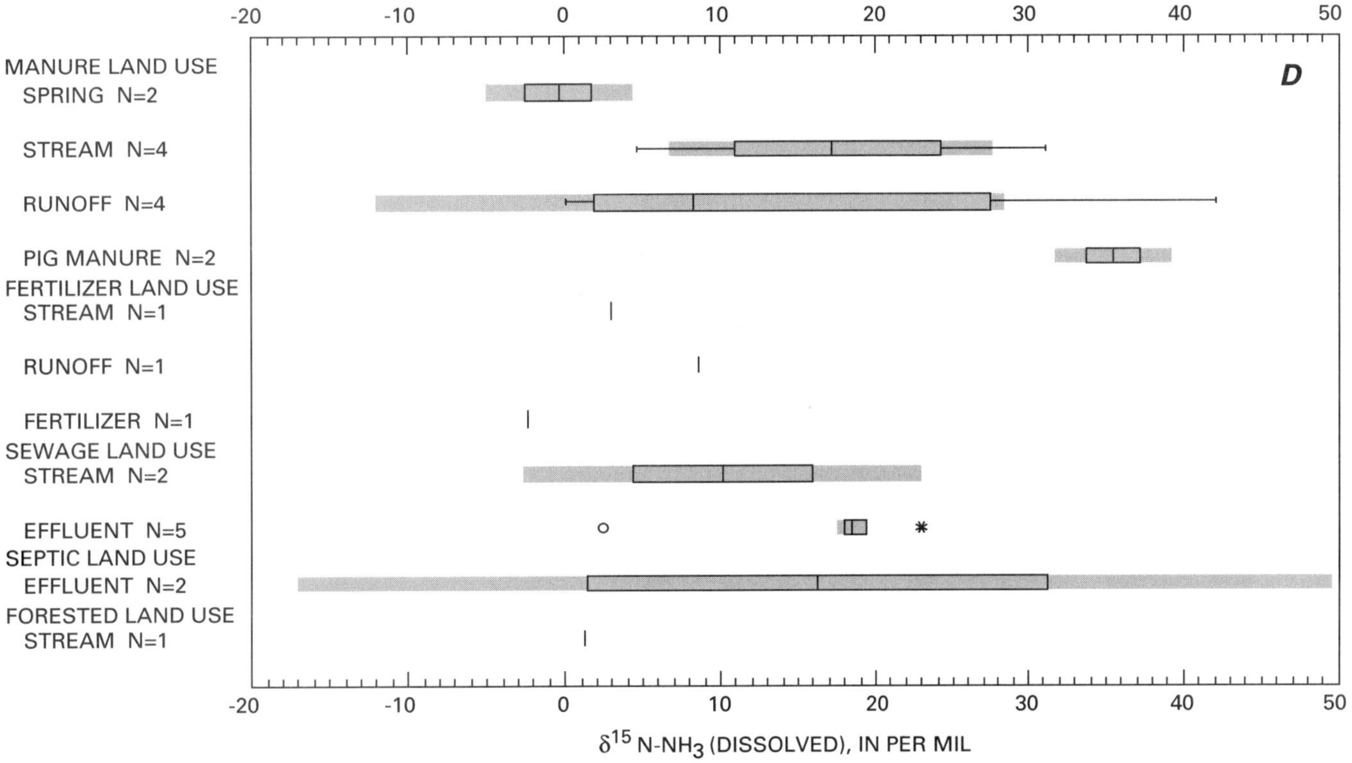

Figure 7. $\delta^{13}C$, $\delta^{15}N$, $\delta^{34}S$, and molar ratio of organic-C to total-N (C-org:N) in N-source, soil, and water samples, Lower Susquehanna River Basin, Pennsylvania—Continued.

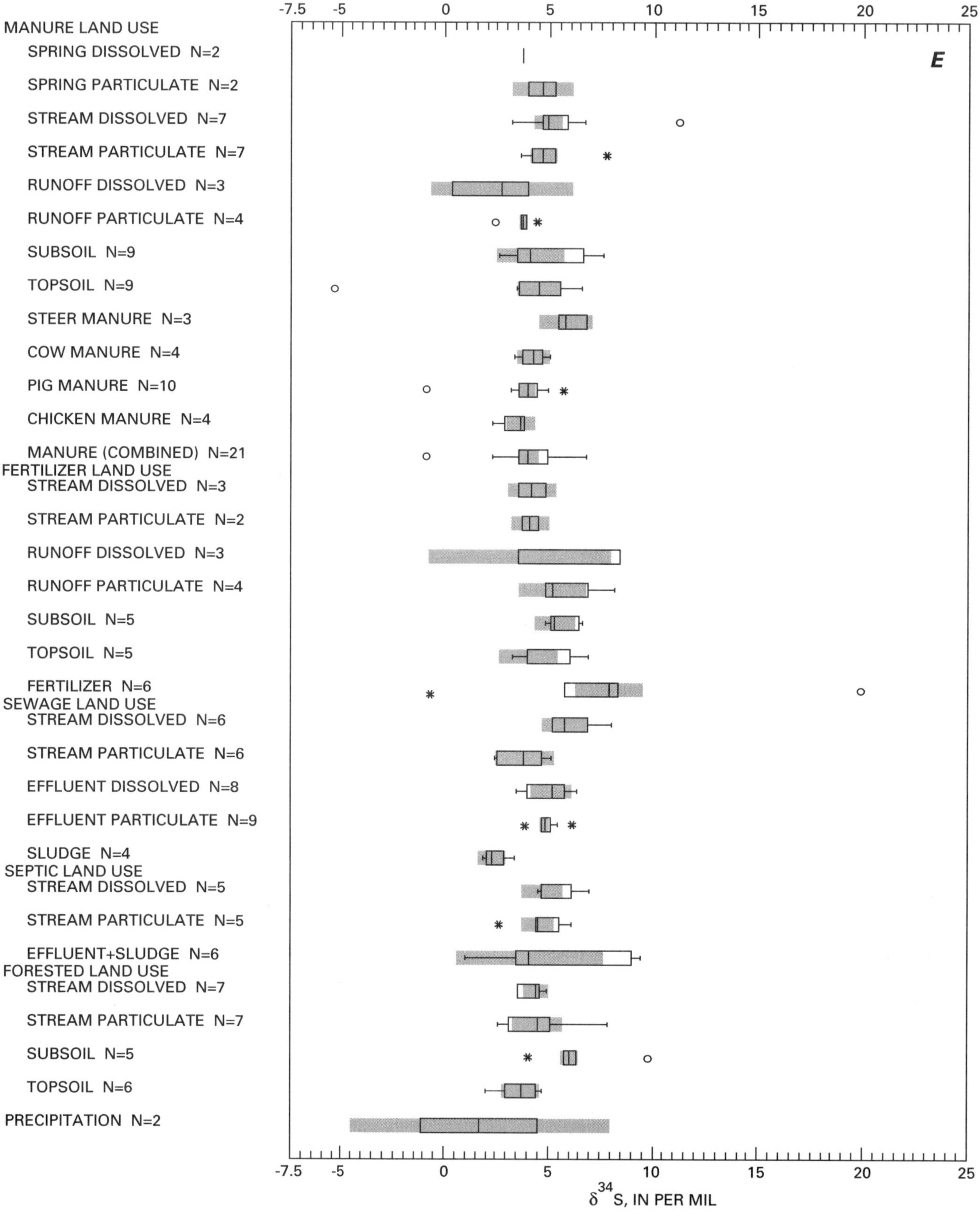

Figure 7. $\delta^{13}C$, $\delta^{15}N$, $\delta^{34}S$, and molar ratio of organic-C to total-N (C-org:N) in N-source, soil, and water samples, Lower Susquehanna River Basin, Pennsylvania—Continued.

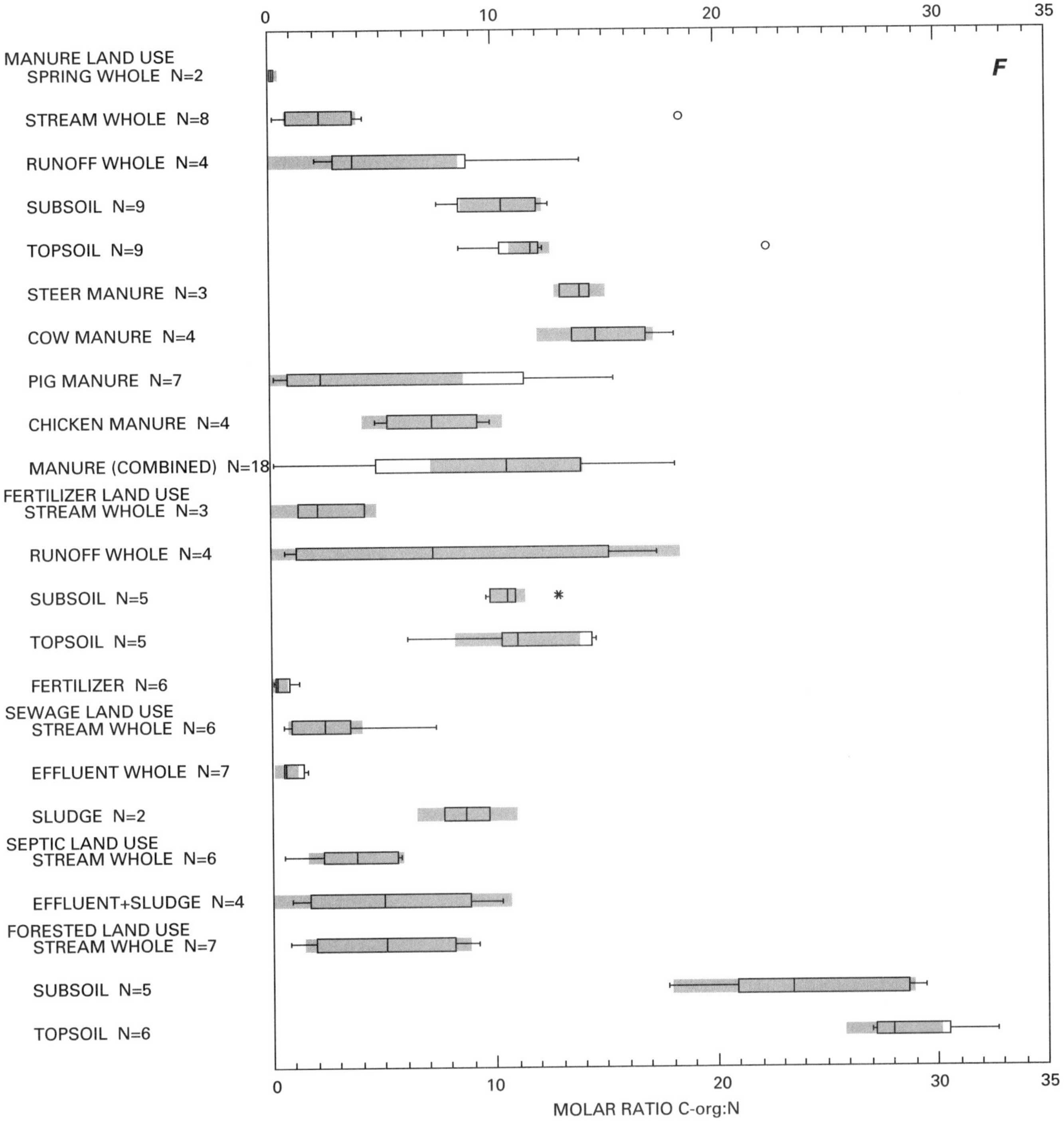

Figure 7. $\delta^{13}C$, $\delta^{15}N$, $\delta^{34}S$, and molar ratio of organic-C to total-N (C-org:N) in N-source, soil, and water samples, Lower Susquehanna River Basin, Pennsylvania—Continued.

tissues (Brady, 1974, p. 534–546). On a unit-weight basis, poultry manure is the richest and cattle manures are the poorest N sources (table 7).

Manure is normally spread on local fields and pastures surrounding the brooding houses, pens, and barns. The spread manure consists of feces, urine, bedding (litter), and feed waste. The chemical composition of this material varies widely depending on animal species, composition of feed, nature and amount of litter, and the handling and storage of manure before it is spread on the land. The predominant feeds of the various farm animals are corn silage, corn, alfalfa, soybean, and grasses. The animals, especially poultry, also may be fed protein supplements of meat-processing by-products and fish meal. Litter compositions generally consist of wheat straw, oat straw, wood chips, and peanut shells. Poultry and steer manures are relatively solid and are commonly stockpiled until spread on fields. Dairy cattle and swine manures are more liquid and commonly are collected in lagoons prior to application.

Concentrations of different N species in manure samples collected for this study are, in order of decreasing concentrations, organic N, NH_3, and NO_3 (fig. 4). Chicken manure (BR1MC, LC2MC) contains the highest concentrations of total N—mostly as organic-N (tables 5 and B3). Swine feces, feeder-cattle, and dairy-cattle manure contain about one-half the concentration of N in chicken manure. The different animal manures contain similar concentrations of total C and total S (fig. 4). Isotopically, however, manure from different animals can be different (fig. 7, table 6) because of differences in animal diet and manure handling. On the basis of $\delta^{13}C$, steer manure (–14.2‰ to –12.2‰) can be distinguished from cow manure (20.6‰ to –18.4‰), swine manure (–18.4‰ to –14.9‰), and chicken manure (–18.8‰ to –17.8‰). On the basis of $\delta^{15}N$, chicken manure generally can be distinguished from cow manure and steer manure (fig. 7); however, ranges of $\delta^{15}N$ overlap among the different classes: chicken manure (1.1‰ to 10.2‰), steer manure (9.5‰ to 11.6‰), cow manure (3.8‰ to 9.1‰), and swine manure (6.3‰ to 37.2‰). Although $\delta^{34}S$ for swine manure is relatively variable (–0.9‰ to 5.8‰), other animal manures have narrower, overlapping ranges of $\delta^{34}S$: chicken manure (2.3‰ to 3.8‰), steer manure (5.4‰ to 6.8‰), and cow manure (3.4‰ to 5.0‰).

Human Septic and Sewage Waste

Human waste generally is disposed through onsite septic systems in rural areas and through wastewater-treatment plants in urban and suburban areas in the study area. Septic effluent normally is piped from an anaerobic holding tank to an onsite leach field, where percolation through the soil promotes the removal of pathogens and nutrients by sorption and denitrification. Periodically, septic sludge is pumped from the tank into vacuum trucks that may dispose of the waste on cultivated fields. Hence, septic effluent and sludge can be nonpoint sources of N contamination. Alternatively, septic-tank waste may be delivered to nearby municipal sewage-treatment plants for processing. In general, municipal sewage treatment is designed to remove or reduce concentrations of biodegradable organic matter, solids, and solutes such as nutrients and metals; neutralize acidity; eliminate odors and bacteria; and saturate the effluent with oxygen. The effluent is processed and then discharged directly as a point source into streams or used in spray irrigation. The sludge can be incinerated, landfilled, or, if contaminant-free, composted and spread on the land surface. A sample of filter-cake sludge from the York wastewater plant was reported to have a nutrient

Table 7. Moisture and nutrient content of manure from farm animals

[Modified from Brady (1974, p. 538); mg/kg, milligram per kilogram]

Animal	Feces/urine ratio	Percentage water	N (mg/kg)	P_2O_5 (mg/kg)	K_2O (mg/kg)
Dairy cattle	80:20	85	5.00	1.35	3.75
Feeder cattle	80:20	85	5.95	2.35	3.55
Poultry	100:0	62	14.95	7.15	3.50
Swine	60:40	85	6.45	3.55	5.45
Sheep	67:33	66	11.50	3.50	10.85
Horse	80:20	66	7.45	2.75	6.60

content as $N\text{-}P_2O_5\text{-}K_2O$ of 6.17-5.09-0.15 (John S. Smith, York Wastewater Treatment Plant, York, Pa., written commun., 1989). However, because sludge from the York plant contains toxic metals—such as zinc, cadmium, and lead—from industrial sources, it cannot be spread on the land surface and must be landfilled.

Septic and sewage-treatment systems commonly produce effluents with different chemical and isotopic characteristics because septic-system processes are anaerobic and sewage-treatment processes are aerobic. However, septic-tank sludge (HW HS in tables B3 and B4) and sewage sludge from the three different sewage-treatment plants (HWOHW, YW HW, DWAHW in tables B3 and B4), which have different treatment processes (Appendix A), are similar chemically. The sludges contain about 20 to 30 wt % total C, 3 to 4 wt % total N, and 0.6 to 0.8 wt % total S (table 5). The septic sludge and sewage sludge differ isotopically, however, with respective means of $\delta^{13}C$ of –24.1‰ and –21.8‰, $\delta^{15}N$ of –2.1‰ and +7.8‰, and $\delta^{34}S$ of 1.0‰ and 2.3‰ (table 6). Furthermore, corresponding particulate and dissolved fractions of effluent from septic or sewage- treatment systems generally have different N and S isotopic ratios (fig. 7) as a result of isotopic fractionation in the anaerobic or aerobic processes.

Although septic-tank samples of sludge plus effluent (HW HS) and septic-field effluent samples (BK2HS) collected for this study have different total concentrations of solutes and different proportions of organic and inorganic species, none of the septic sludge or effluent samples contain detectable concentrations of NO_3-N (table B3). Septic-tank effluent is more concentrated than septic-field effluent. Septic-tank effluent contains predominantly organic C, organic N, and SO_4^{2-}; septic-field effluent contains predominantly inorganic C, ammonium-N, and SO_4^{2-}. Dissolved and particulate fractions of the septic effluents have $\delta^{13}C$ of –23.4‰ to –21.9‰, $\delta^{15}N$ of –0.7‰ to +31.2‰, and $\delta^{34}S$ of 3.5‰ to 9.4‰ (table 6). C, N, and S in the particulate fraction of septic-tank effluent are isotopically heavier than in the sludge, and N and S in the particulate fraction are isotopically lighter than in the corresponding dissolved fraction (figs. 6 and 7).

Sewage effluents from the York and Dillsburg treatment plants are more oxidized than that from the Harrisburg plant. Effluent from the York and Dillsburg plants contains less than 10 mg/L total N, which is at least 80 wt % NO_3-N, and that from the Harrisburg plant contains more than 20 mg/L total N, which is at least 70 wt % NH_3-N (fig. 4 and table B3). However, isotopically the effluents from the three plants are similar. Sewage effluent particulate has $\delta^{13}C$ of –24.3‰ to –20.8‰, $\delta^{15}N$ of –3.7‰ to +10.2‰, and $\delta^{34}S$ of 3.6‰ to 6.1‰ (table 6). Relative to the particulate, the dissolved fraction of sewage effluent has isotopically heavier N, with values of $\delta^{15}N$-inorg of 13.7‰ to 26.5‰ (figs. 6 and 7). However, values of $\delta^{34}S$ for the particulate and dissolved fractions of sewage effluent are similar (figs. 6 and 7).

Isotopic Differences Among Nitrogen Sources

The wide ranges of $\delta^{13}C$, $\delta^{15}N$, and $\delta^{34}S$ values for comparable N-source materials (table 6) indicate that there are significant isotopic variations within each N-source class. Multiple notched boxplots (fig. 7) were compared to evaluate significant differences among medians for different classes. If the notched intervals about medians for different classes do not overlap, the medians are significantly different at the 95-percent confidence level (Velleman and Hoaglin, 1981).

On the basis of $\delta^{13}C$ (fig. 7*A*), three classes of N sources can be distinguished from one another: forest litter, human waste (septic + sewage), and animal manure, in order of increasing ^{13}C-enrichment. Although synthetic fertilizer $\delta^{13}C$ values overlap with those of human waste and forest litter, they are distinctly different from animal manure. On the basis of the C-org:N ratio (fig. 7*F*), forest soils (C-org:N of 18 to 33) potentially can be distinguished from anthropogenic N sources and cultivated soils (C-org:N of less than 1 to 18).

N-source samples cannot be distinguished from one another on the basis of $\delta^{15}N$ and $\delta^{34}S$ (figs. 7*B*–7*E*). Wide variations of $\delta^{15}N$ within human-waste (sewage and septic) and animal-manure classes and similarity of $\delta^{15}N$ for fertilizer and forest sources make $\delta^{15}N$ values alone of limited use for determining N source. The small overall range of 20‰ for $\delta^{34}S$ diminishes the use of this measure because most of the N sources have overlapping compositions. Only forest litter (topsoil) and synthetic fertilizer have significantly different median $\delta^{34}S$ values.

Characterization of Soil and Water Samples from Different Land-Use Areas

In general, $\delta^{13}C$, $\delta^{15}N$, and $\delta^{34}S$ values for many corresponding N-source, soil, and surface-water samples are significantly different within a land-use category, and those for surface-water samples overlap among categories (fig. 7). Hence, comparisons of single isotopes are of limited value for defining N sources to streams.

Evaluation of combined $\delta^{13}C$, $\delta^{15}N$, and $\delta^{34}S$ data could be helpful to indicate characteristics of comparable materials, sources of the elements, and processes affecting element cycling in the different land-use areas (Rau and others, 1981; Spiker and Kendall, 1983; Peterson and others, 1985, 1986; Peterson and Fry, 1987). Bivariate plots of $\delta^{13}C$, $\delta^{15}N$, $\delta^{34}S$, and C-org:N (figs. 8–11) were evaluated to help determine isotopic characteristics of corresponding N-source, soil, and surface-water-particulate samples from different land-use areas. In figures 8–11, plots on the left show individual data points for different land-use classes and plots on the right show rectangles that enclose 70 percent of these data as defined by the 15th- to 85th-percentile values.

Figure 8 shows that high values of $\delta^{13}C$ and $\delta^{15}N$ in soils and surface waters are characteristic of manure-use areas, intermediate values are characteristic of fertilizer-use areas, and lower values are related to forested and(or) septic land-use areas. Figure 9 shows similar groupings for soil data on the basis of $\delta^{13}C$ relative to $\delta^{34}S$. Figures 10 and 11 show that values of C-org:N relative to $\delta^{13}C$ and $\delta^{15}N$ aid in distinguishing between agricultural and forest soils.

Corresponding surface-water data shown in figures 8–11 do not reveal characteristic data groupings for different land-use classes. The failure of surface-water data to cluster into different land-use groups relative to those of associated N-sources or soils (figs. 8–11) and the significant differences between isotopic compositions of N-source and associated surface-water samples (fig. 7) indicate that C, N, and S compounds are fractionated during transport and that additional sources or sinks of the elements may be present along transport paths.

Relatively uniform C, N, and S isotopic compositions of stream waters from different land-use areas (figs. 8–11) diminish the use of the isotopic data for determining the N sources in the stream waters but could be useful for other applications. For example, the streams contribute to C, N, and S loads transported from the Susquehanna River Basin to the Chesapeake Bay. The stable isotope compositions of the transported terrigenous materials could be used to determine organic matter contributions from terrigenous, estuarine, and marine sources to food webs in the bay or other estuaries (see Peterson and others, 1985, 1986; Peterson and Fry, 1987).

Forest: Stony Creek and Upper Dogwood Run

Concentrations of C, N, and S in forested-area samples from the Stony Creek (SC) and Dogwood Run (D1) Basins (fig. 1) generally decrease in the order of organic-rich topsoil, underlying mineral subsoil, and nearby stream waters (table 5). Stream-water samples from the forested areas contained total C from 2.4 to 8.3 mg/L, total N from 0.6 to 2.1 mg/L, and total S from 0.4 to 4.4 mg/L.

Minimal isotopic fractionation and narrow ranges of $\delta^{13}C$ and $\delta^{15}N$ are expected for soil and water of undisturbed forested areas because leaf litter is the only major source of C and N and the N supply is efficiently recycled. Medians of $\delta^{13}C$ for forest topsoil, subsoil, and stream-water particulates are not significantly different (fig. 7*A*). However, an apparent increase in median $\delta^{13}C$ from –27.3‰ to –26.5‰ from the topsoil to underlying subsoil and an apparent decrease in C-org:N from 28 to 23 (fig. 7*F*) suggest potential isotopic fractionation associated with decomposition and CO_2 loss (Nadelhoffer and Fry, 1988). Large differences in median C-org:N between forest soils and stream water (fig. 7*F*) indicate that the particulates in the stream are not simply smaller fragments of soil-derived material. It is likely that in-stream decomposition has caused relative losses of C and gains of N associated with colonization by microbes (Berg and Staaf, 1981; Fairchild and others, 1983; James and others, 1988).

Median $\delta^{15}N$ values are 0.5‰ for topsoil and 4.5‰ for subsoil in the forested areas (fig. 7*B*). The difference in $\delta^{15}N$ between soil horizons can be attributed to sorption of ^{15}N-enriched ammonium in the subsoil and to the preferential assimilation of ^{14}N by microbes in the topsoil. A similar difference in composition between the particulate and dissolved N fractions in stream-water samples probably results from microbial activity and subsequent fractionation in the stream. However, there may be other explanations. For example, similar $\delta^{15}N$ values for topsoil and

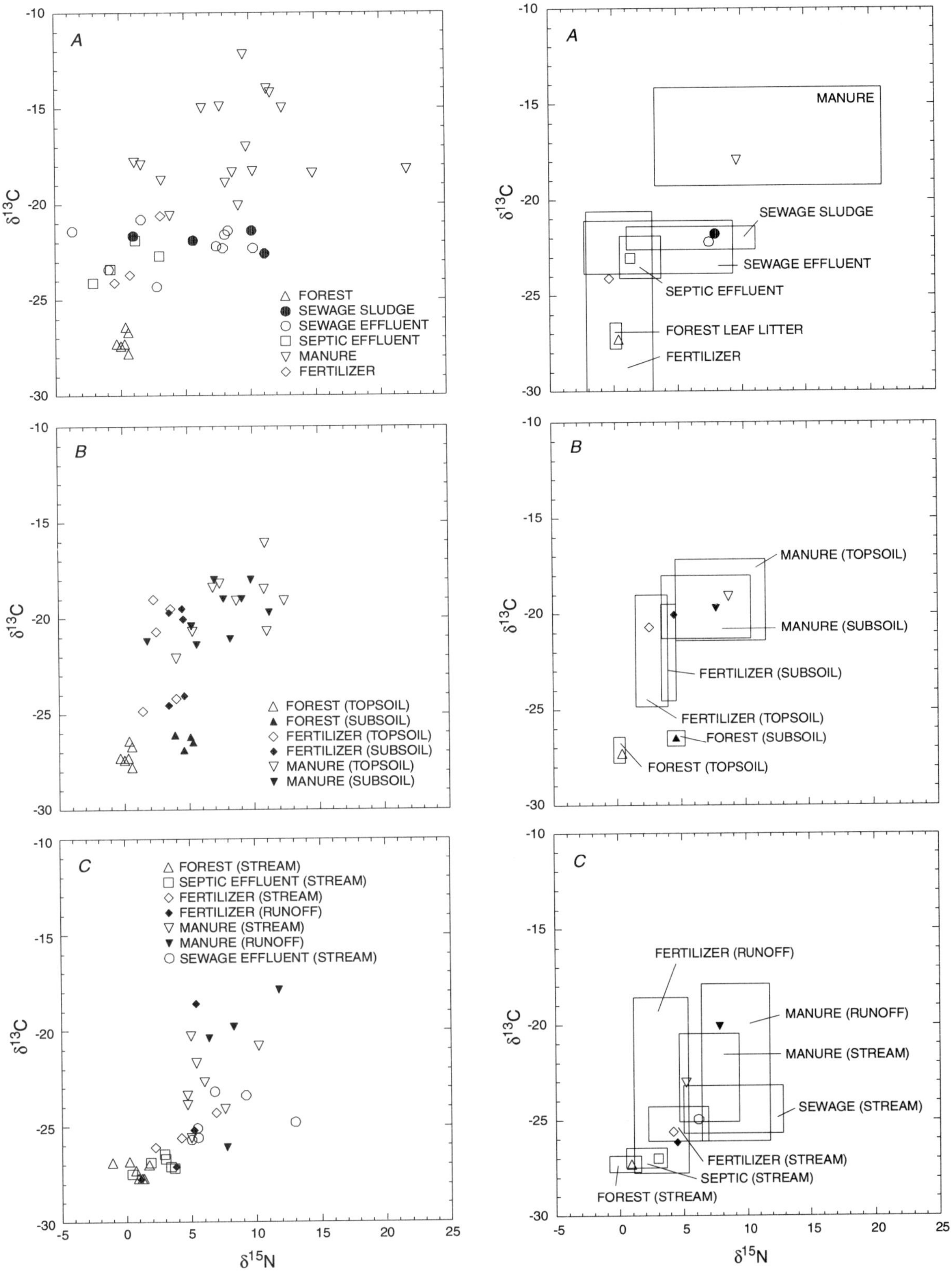

Figure 8. $\delta^{13}C$ relative to $\delta^{15}N$ for: *A*, N-source materials; *B*, soil; and *C*, suspended particulates in water, Lower Susquehanna River Basin, Pennsylvania. (Plots on the left show individual data points, and plots on the right show median within 15th- to 85th-percentile range for data classes.)

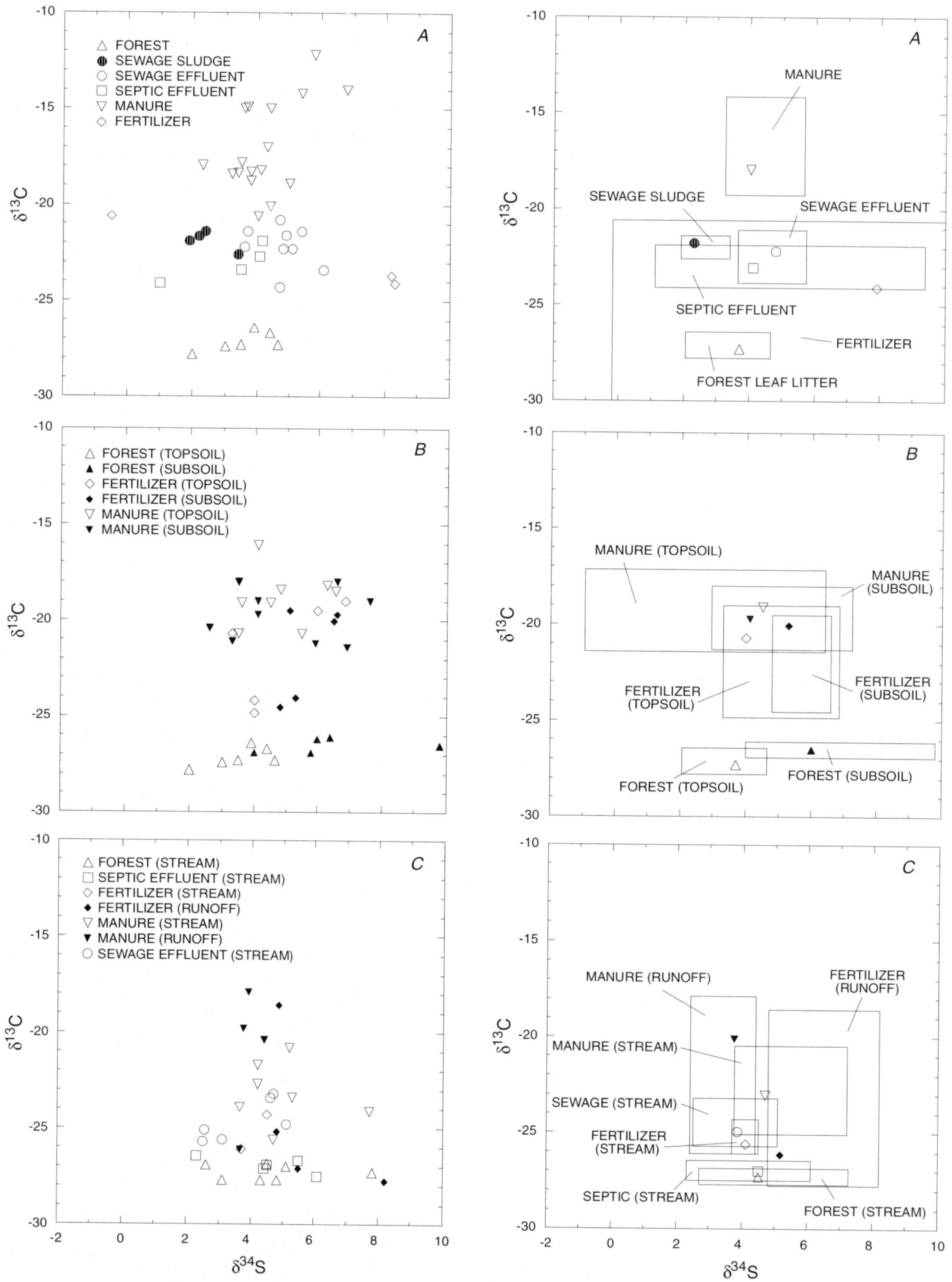

Figure 9. $\delta^{13}C$ relative to $\delta^{34}S$ for: *A*, N-source materials; *B*, soil; and *C*, suspended particulates in water, Lower Susquehanna River Basin, Pennsylvania. (Plots on the left show individual data points, and plots on the right show median within 15th- to 85th-percentile range for data classes.)

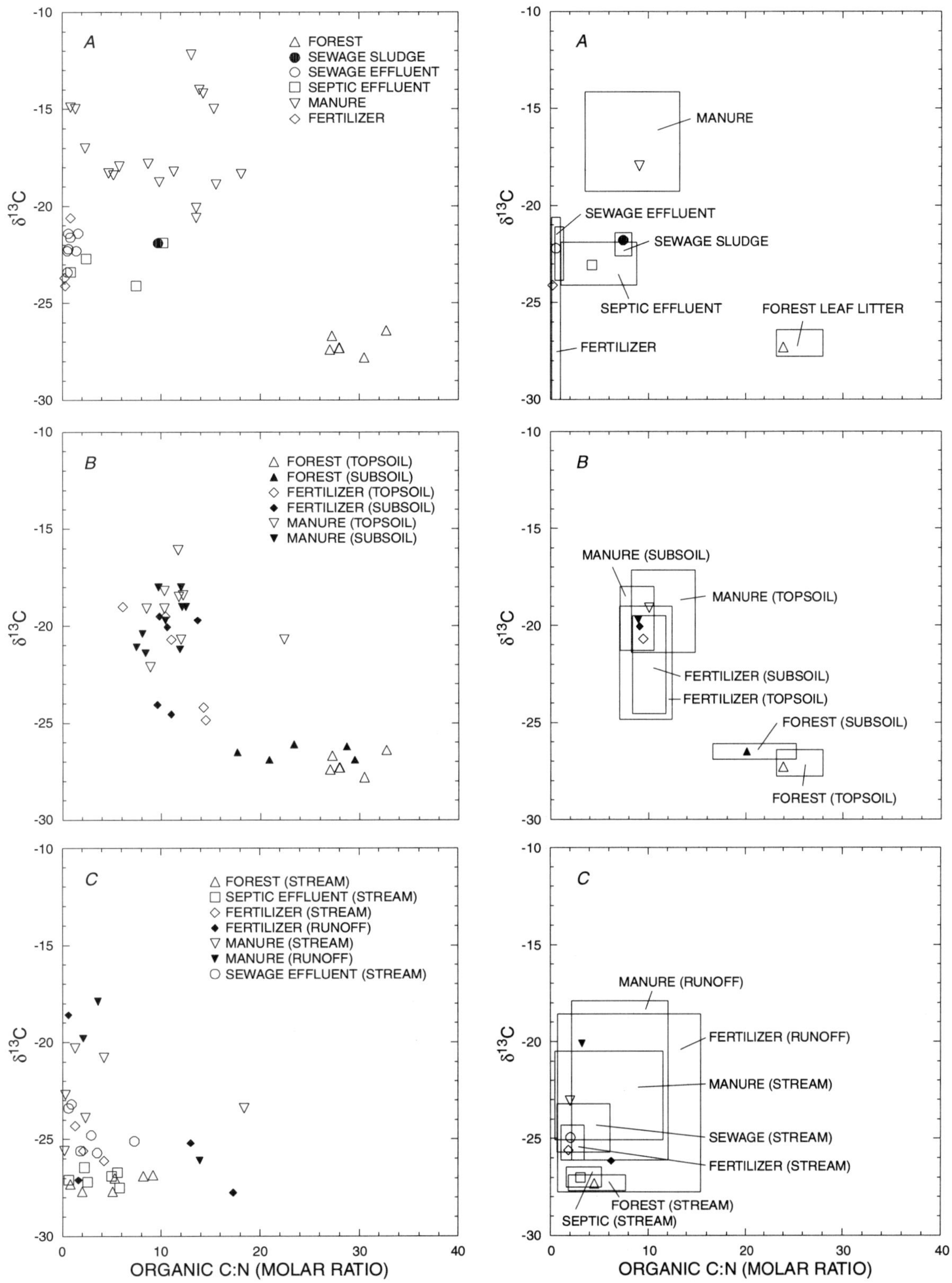

Figure 10. $\delta^{13}C$ relative to C-org:N for: *A*, N-source materials; *B*, soil; and *C*, suspended particulates in water, Lower Susquehanna River Basin, Pennsylvania. (Plots on the left show individual data points, and plots on the right show median within 15th- to 85th-percentile range for data classes.)

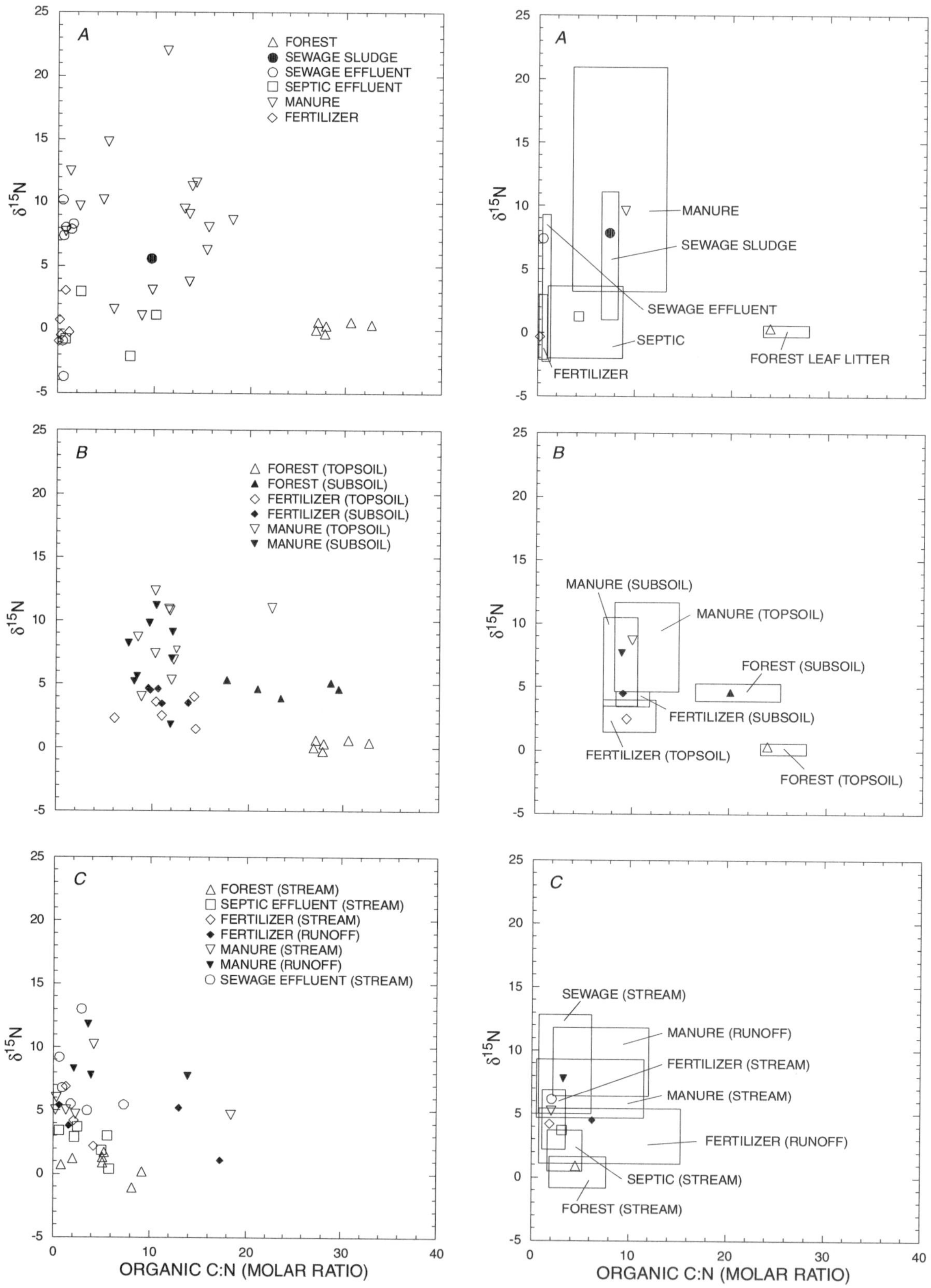

Figure 11. $\delta^{15}N$ relative to C-org:N for: *A*, N-source materials; *B*, soil; and *C*, suspended particulates in water, Lower Susquehanna River Basin, Pennsylvania. (Plots on the left show individual data points, and plots on the right show median within 15th- to 85th-percentile range for data classes.)

particulate in stream water draining the forested watershed can result if the particulate consists simply of eroded leaf litter. Also, similar $\delta^{15}N$ values for subsoil and dissolved N in the stream water can result from leaching of mineralized soil N.

Smaller median $\delta^{34}S$ values in the forest topsoil (3.7‰) compared to the subsoil (6.0‰) (fig. 7*E*) may result from preferential assimilation of ^{32}S by microbes in the topsoil. Median $\delta^{34}S$ of the particulate and dissolved S fractions in stream-water samples (4.5‰ and 4.4‰, respectively) are similar to one another (fig. 7*E*) and can result from mixing of S from the topsoil and subsoil horizons. Hence, although stable isotopes and C-org:N signatures appear to be different in topsoil and subsoil of forested basins, use of these measures to trace contributions of C, N, and S to the stream water is complicated by isotopic fractionation during transport and by the need to account for differences in the compositions of dissolved and suspended fractions in the stream water.

Fertilizer: Monroe Creek and Bald Eagle Creek

Unfiltered runoff-water samples from the synthetic-fertilizer-use areas, including a golf course in the Monroe Creek (M2A) Basin and an agricultural field site in the Bald Eagle Creek (BE1) Basin (fig. 1), had concentrations of total C of 20 to 28 mg/L, total N of 0.9 to 13.3 mg/L, and total S of 1.0 to 2.4 mg/L (table 5).

Medians of $\delta^{13}C$ for fertilizer, topsoil, subsoil, and runoff-water particulates, which range from –26.2‰ to –20.1‰, are not significantly different (fig. 7*A*). The lack of a difference is due, in part, to combining data for the golf course and farm sites. Values of $\delta^{13}C$ for soil samples from each of the two land-use areas have narrow ranges and significantly different medians, reflecting different soil-C sources, namely turf grasses at the golf course and mixed corn, wheat, and soybean crops at the farm field. Values of $\delta^{13}C$ for soil samples from the golf course range from –25.3‰ to –24.0‰ and those from the farm field range from 20.7‰ to –19.0‰ (table B4). Particulate-C in runoff-water samples from each site have more variable compositions. Particulates in two runoff-water samples from the golf course had $\delta^{13}C$ of –25.2‰ and 27.8‰, compared with –18.6‰ and –27.1‰ in two runoff samples from the farm field. The particulate ^{13}C in runoff is similar to that in stream waters. Values of $\delta^{13}C$ for stream-water particulates are lower than those for associated fertilized soils, probably because of contributions by algae and leaf litter from trees growing upstream and along the streambank. Hence, the C-isotopic compositions of particulates in runoff and stream-water samples from the fertilizer-use areas do not clearly reflect $\delta^{13}C$ values of fertilizer or soil but some combination of these and possibly other sources. Because the mass of C in annual applications of synthetic fertilizer commonly is small compared to that in soil humus and leaf litter, synthetic fertilizer probably is only a minor source of C in runoff and stream water.

The fertilizer, topsoil, and subsoil samples generally show successive enrichment in ^{15}N; the median $\delta^{15}N$ values are –0.3‰, 2.5‰, and 4.5‰, respectively (table 6). Topsoil samples in the fertilizer-use subbasins have median $\delta^{15}N$ values significantly larger than those for fertilizer and forest topsoil but similar to those for forest subsoil (fig. 7*B*). The relative enrichment of ^{15}N in the topsoil can result from mixing with subsoil by plowing or aeration practices and from volatilization of ^{15}N-depleted NH_3 after fertilizer N is applied. The 2‰ difference between medians of $\delta^{15}N$ for topsoil and subsoil horizons is less than that for the forested watershed but is significant and can result from the same N-cycle processes described for an undisturbed forested soil.

The N-isotopic composition of fertilized-soil samples is reflected in the runoff waters draining the fertilizer-use subbasins. Median $\delta^{15}N$ values for the dissolved and particulate fractions in runoff water from the fertilizer-use areas are statistically indistinguishable from those of associated topsoil and subsoil. However, values of $\delta^{15}N$-NO_3 for stream water collected at a downstream location (M2AWS in table B4) and at an upstream location (M1 WS in table B4) in the fertilizer-use watershed are similar to one another and are larger than $\delta^{15}N$ values for fertilizer and fertilized soils. Hence, the "fertilizer-use" stream-water composition reflects a mixture of N sources, including fertilizer and ^{15}N-enriched materials from upstream or other areas.

The S-isotopic composition of fertilizer is widely variable. The median $\delta^{34}S$ for fertilizer is higher than those for associated topsoil, subsoil, and dissolved SO_4^{2-} in runoff water and stream water, which are not significantly different from one another (fig. 7*E*). Wide variations of $\delta^{34}S$ values for fertilizer and rainfall samples relative to those for soil and water samples (fig. 7*E*) may indicate that a relatively constant S supply, such as ground-water SO_4^{2-} from mineral dissolution, is the predominant source of S in the surface-water samples.

The similarity of $\delta^{13}C$, $\delta^{15}N$, and $\delta^{34}S$ values for topsoil and particulates in surface water of the fertilizer-use areas, as in forested areas, may result if particulate matter in streams consists mainly of eroded topsoil. This argument is supported by the lack of a significant difference in C-org:N for the topsoil and runoff water in the fertilizer-use areas (fig. 7*F*). However, stream C-org:N is lower than that for topsoil (fig. 7*F*) and the dissolved NO_3^- in the stream water is relatively enriched in ^{15}N compared to fertilizer and runoff (fig. 7*C*). This indicates that effects from leaching of mineralized soil N and in-stream fractionation processes, and also mixing with ^{15}N-enriched sources such as animal manure (waterfowl) or human waste (sludge by-products), can be important controls of isotopic compositions of stream water in the fertilizer-use subbasins.

Manure: Brush Run and Conestoga River Field Sites

Manure, topsoil, subsoil, runoff water, and stream-water samples from manure-use areas were collected from three agricultural field sites—one in the Brush Run (BR1) Basin and two in the Conestoga River (C1, C2) Basin (fig. 1). Runoff-water samples from these areas had extremely variable compositions; concentrations of total C ranged from 26 to 533 mg/L, total N ranged from 1.7 to 122 mg/L, and total S ranged from 1.0 to 20 mg/L (table 5).

Manure and associated topsoil, subsoil, and particulate fractions in runoff water, stream water, and ground water are enriched in ^{13}C and ^{15}N relative to equivalent materials from forested land-use areas and have widely variable, overlapping, isotopic compositions (figs. 7*A*–7*D*). As discussed previously, manure from different animals can have different $\delta^{13}C$ and $\delta^{15}N$, depending on the animal species, its diet, and manure-handling practices. However, for comparison with associated soils and waters, an overall group of manure is shown in figure 7, because manure from a variety of animals is applied to the fields studied. Medians of $\delta^{13}C$ and $\delta^{15}N$ for manure and corresponding topsoil, subsoil, and runoff-water particulate are not significantly different (fig. 7*A*). The similarity in $\delta^{13}C$ and $\delta^{15}N$ in manure, topsoil, shallow subsoil, and runoff particulate results from recycling of locally grown fodder-plant materials (corn, hay, alfalfa) in manure plus plowing of the fields, which tends to homogenize the soil. Plowing promotes sediment erosion, which contributes to the particulate fraction in runoff. In contrast, particulates in nearby stream waters generally are depleted in the heavier C and N isotopes.

Water samples collected from manure-use study sites also are characterized by wide ranges in $\delta^{15}N$ for the dissolved fractions of runoff, stream, and spring waters, which are statistically indistinguishable from those of the corresponding particulate, topsoil, or original manure sources (fig. 7*B*). Medians of $\delta^{15}N$ for stream water NO_3-N are intermediate to those for runoff and spring waters, and thus are consistent with those expected from mixing of the runoff and ground waters.

Although steer manure is slightly enriched in ^{34}S, different animal manures have similar median $\delta^{34}S$. The $\delta^{34}S$ for associated topsoil, subsoil, and runoff-water particulate and dissolved S are indistinguishable from manure and from one another (fig. 7*E*). However, the soil and water in manure-use areas have a relatively wide range of $\delta^{34}S$ from –5.4‰ to +13.0‰ compared with a range of $\delta^{34}S$ from –0.9‰ to +6.8‰ for manure, which indicates other S sources and fractionation during transport could be significant factors affecting the $\delta^{34}S$ of soil and water.

Although manure, soil, and water from manure-use areas have statistically similar isotopic compositions, because of the wide range in $\delta^{13}C$, $\delta^{15}N$, and $\delta^{34}S$ and overlap with other N sources, the isotopic data from streams could not be used to identify manure as the primary N source in the water. The wide variability of C, N, and S isotope data for soil and water from the manure-use areas indicates that fractionation during transport of C, N, and S is likely and that additional sources of N and S, such as rainfall and soil minerals, respectively, could be important.

Septic: Middle Dogwood Run and Berkshire Hills

Stream-water samples, which were presumed to be affected by septic systems, were collected from two rural residential areas in the Dogwood Run Basin. One sampling location was along the middle reaches of Dogwood Run (D2A) and the other was along an unnamed tributary near its mouth in the Berkshire Hills (BH2) development (fig. 1). These represent downstream water-quality samples relative to the forest water-quality data described previously. Stream waters draining septic land-use areas had concentrations of total C ranging from 4.9 to 47 mg/L, total N ranging from 0.5 to 2.8 mg/L, and total S ranging from 1.0 to 9.0 mg/L (table 5). Relative to comparable upstream waters (table 5), which had concentrations of total C ranging from 2.4 to 8.3 mg/L, total N ranging from 0.6 to 2.1 mg/L, and total S ranging from 0.4 to 4.35 mg/L, concentrations of total C and S appear to be greater by about a factor of 2, but concentration of total N is relatively unchanged. Hence, in the subbasins studied, septic effluent may not contribute measurable concentrations or loads of N in downstream samples.

Denitrification and assimilation of N along transport paths from septic fields to downstream points and within streambeds can reduce concentrations of N and produce ^{15}N-enrichment in downstream samples. However, upstream and downstream waters in the septic-use areas have indistinguishable isotopic compositions and C-org:N ratios (fig. 7).

Sewage: Lower Dogwood Run and Codorus Creek

Sewage effluent was collected at three sewage outfalls, and stream water was collected upstream and downstream from two of these outfalls. One outfall is along the lower reaches of Dogwood Run near the Dillsburg sewage plant (DWA) and one is along Codorus Creek near the York sewage plant (YW) (tables A1, B2–B4). Downstream waters were collected at locations D4 and CCP for comparison with upstream waters at locations D3 and CCY, respectively (fig. 1). Concentrations of total C, total N, and total S were greater in sewage effluent than they were in downstream water. Hence, elevated concentrations of C, N, and S in downstream water relative to upstream water indicate that sewage effluent contributes to the solute concentrations and loads in downstream water.

Sewage effluent and downstream particulates from the above locations have similar median δ^{15}N and δ^{34}S and dissimilar median δ^{13}C (fig. 7). Nevertheless, upstream and downstream waters have statistically indistinguishable isotopic compositions. Although the sewage effluent was slightly enriched in ^{13}C compared to the upstream water, in-stream loads of C were much greater than the contributions from sewage effluent (table 8) causing dilution of the sewage isotopic signature.

ESTIMATION OF NITROGEN LOADS IN STREAM WATERS

Isotopic mass balance offers potential for estimating nutrient loads from different sources that can be identified on the basis of their isotopic compositions. Hence, data collected for this study were used to estimate loads. Before discussing results of mass-balance computations, computation methods and the format of table 8 are described below.

Measured δ^{13}C, δ^{15}N, and δ^{34}S for upstream water, downstream water, and the primary N source contributing nutrient loads in the stream reach were used in equations 2, 3, 4, and 5 to estimate loads of C, N, and S that could be derived from the N source. These measured isotope delta values are shown in columns 3, 5, and 7 in table 8. Measured minimum and maximum isotope delta values for the N source also are shown in table 8, in columns 8 and 9, for comparison with computed estimates of δ^{13}C, δ^{15}N, and δ^{34}S for the N source, in column 10. For water and aqueous N-sources, the average isotopic composition of total dissolved N, δ^{15}N-inorg, was used in the computations. Mean values for N-source δ^{13}C, δ^{15}N, and δ^{34}S and concentrations of C, N, and S were used if data concurrent with surface-water samples were not available. Computed loads for upstream and downstream waters are shown in columns 4 and 6, respectively, for each sampling date. The difference in these loads, which corresponds to the amount of C, N, or S added from the N source, is shown in column 12. Negative values are shown as missing in column 12 because losses of flow or C, N, and S over the short reaches between sampling points were not likely to be significant—distances between upstream and downstream sampling points were generally only tens of meters and were always less than 3 km. In addition, three comparable values for N-source chemical loads computed by different methods are shown in table 8, in columns 10, 13, and 14. Ideally, the estimated isotopic composition of the N source (column 10) should be equal to the measured composition (column 7) and should be within the range of measured values (columns 8 and 9). Also, the load computed as the difference between concurrent downstream and upstream loads (column 12) should be equal to that computed directly from discharge rate and concentration of the N source (column 10) and those computed indirectly by the isotopic mass difference between downstream and upstream loads

Table 8. Chemical loads and isotopic compositions of nitrogen sources estimated by mass balance in the Lower Susquehanna River Basin, Pennsylvania

[‰, per mil; kg/d, kilogram per day; mg/L, milligram per liter; ML/d, million liters per day; --, no data]

1	2	3	4	5	6	7	8	9	10	11	12	13	14
		Upstream[1]		Downstream[1]		N-source isotope composition[2]				Estimates of chemical load from N source			
												Isotope mass balance	
Surface-water constituent	Date sampled	Delta, δE_{up}[2] (‰)	Load, QE_{up}[3] (kg/d)	Delta, δE_{dn}[2] (‰)	Load, QE_{dn}[3] (kg/d)	Meas., δE_{NS} (‰)	Min., δE_{min} (‰)	Max., δE_{max} (‰)	Est., δE_{est} (‰)	Direct, QE_{NS}[1, 3] (kg/d)	Difference, QE_{dn}-QE_{up}[4] (kg/d)	NSQ.I[4] (kg/d)	NSX.I[4] (kg/d)
					Dogwood Run (D3 to D4): Sewage (DWA)								
C, particulate	881215	-29.30	87.54	-23.40	110.46	-22.20	-22.60	-21.60	-0.87	37.10	22.92	0.89	91.79
	890518	-27.50	1024.40	-25.60	1642.45	-22.20	-22.60	-21.60	-22.45	--	618.05	625.03	588.80
	890706	-26.85	1015.83	-25.70	1304.78	-22.20	-22.60	-21.60	-21.66	--	288.95	281.88	322.69
	891012	-27.40	144.13	-23.20	214.64	-22.20	-22.60	-21.60	-14.61	37.10	70.51	46.42	173.36
N, particulate	881215	3.80	6.51	9.20	20.92	8.93	7.40	11.10	11.64	12.57	14.41	-1.09	22.02
	890518	.70	93.13	5.50	229.39	8.93	7.40	11.10	8.78	--	136.27	22.61	133.79
	890706	2.60	94.25	5.00	104.38	8.93	7.40	11.10	27.33	--	10.13	-4.57	39.58
	891012	4.50	9.16	6.80	25.25	8.93	7.40	11.10	8.11	12.57	16.09	1.92	13.11
N, dissolved	881215	6.90	2.93	15.45	20.50	26.55	26.55	26.55	16.88	11.81	17.57	11.17	8.92
	890518	5.20	93.13	6.00	192.69	26.55	26.55	26.55	6.75	--	99.56	25.31	7.22
	890706	4.90	78.54	--	104.38	26.55	26.55	26.55	--	--	25.84	--	--
	891012	--	9.16	--	22.73	26.55	26.55	26.55	--	11.81	13.57	--	--
S, particulate	881215	4.30	23.11	4.60	54.81	4.36	3.40	5.10	4.82	34.09	31.71	-17.01	274.06
	890518	2.50	592.63	3.10	642.30	4.36	3.40	5.10	10.26	--	49.67	-33.98	207.19
	890706	4.80	298.47	2.50	405.93	4.36	3.40	5.10	-3.89	--	107.47	33.25	2121.92
	891012	4.60	30.37	4.70	63.13	4.36	3.40	5.10	4.79	34.09	32.76	-.16	-26.30
S, dissolved	881215	8.70	5.86	5.30	54.81	4.20	3.50	4.60	4.89	34.02	48.95	57.03	41.41
	890518	8.80	533.36	6.30	642.30	4.20	3.50	4.60	-5.94	--	108.93	-154.08	349.07
	890706	10.30	298.47	8.00	347.94	4.20	3.50	4.60	-5.88	--	49.48	-69.21	131.19
	891012	9.60	28.92	6.90	61.87	4.20	3.50	4.60	4.53	34.02	32.94	35.53	30.93
					Dogwood Run (D2A to D3): Tributary (D2B)								
C, particulate	881215	-26.70	9.12	-29.30	87.54	-	-27.90	-27.10	-29.60	--	78.42	--	--
	890518	-27.50	427.81	-27.50	1024.40	-27.60	-27.90	-27.10	-27.50	589.34	596.59	594.43	0
	890706	-26.90	271.31	-26.85	1015.83	-27.10	-27.90	-27.10	-26.83	618.66	744.53	737.16	-253.96
	891012	-26.45	22.49	-27.40	144.13	-27.90	-27.90	-27.10	-27.58	69.25	121.64	120.22	94.43
N, particulate	881215	3.00	.93	3.80	6.51	--	2.30	4.50	3.93	--	5.58	--	--
	890518	.40	55.20	.70	93.13	2.30	2.30	4.50	1.14	40.97	37.93	-3.60	14.70
	890706	1.85	30.15	2.60	94.25	3.85	2.30	4.50	2.95	32.41	64.11	5.18	35.34
	891012	2.90	2.97	4.50	9.16	4.50	2.30	4.50	5.27	4.91	6.19	-.48	9.16
N, dissolved	881215	3.90	.93	6.90	2.93	--	5.40	5.85	8.29	--	2.00	--	--
	890518	--	34.50	5.20	93.13	5.40	5.40	5.85	--	39.39	58.63	--	--
	890706	4.20	18.84	4.90	78.54	5.85	5.40	5.85	5.12	30.93	59.70	52.26	33.32
	891012	--	2.22	--	9.16	--	5.40	5.85	--	4.65	6.93	--	--
S, particulate	881215	5.50	1.68	4.30	23.11	--	5.50	13.00	4.21	--	21.43	--	--
	890518	6.10	324.31	2.50	592.63	13.00	5.50	13.00	-1.85	288.37	268.32	11.40	-309.20
	890706	4.50	67.83	4.80	298.47	5.60	5.50	13.00	4.89	173.81	230.64	0	81.40
	891012	2.30	5.68	4.60	30.37	5.50	5.50	13.00	5.13	15.50	24.68	.90	21.83
S, dissolved	881215	4.50	1.68	8.70	5.86	--	10.30	12.70	10.39	--	4.18	--	--
	890518	--	324.31	8.80	533.36	12.70	10.30	12.70	--	288.37	209.06	--	--
	890706	6.10	67.83	10.30	298.47	12.60	10.30	12.70	11.54	173.81	230.64	211.15	192.86
	891012	7.00	4.94	9.60	28.92	10.30	10.30	12.70	10.14	15.50	23.98	23.60	22.79

Table 8. Chemical loads and isotopic compositions of nitrogen sources estimated by mass balance in the Lower Susquehanna River Basin, Pennsylvania—Continued

[‰, per mil; kg/d, kilogram per day; mg/L, milligram per liter; ML/d, million liters per day; --, no data]

1	2	3	4	5	6	7	8	9	10	11	12	13	14
		Upstream[1]		Downstream[1]		N-source isotope composition[2]				Estimates of chemical load from N source			
												Isotope mass balance	
Surface-water constituent	Date sampled	Delta, δE_{up}[2] (‰)	Load, QE_{up}[3] (kg/d)	Delta, δE_{dn}[2] (‰)	Load, QE_{dn}[3] (kg/d)	Meas., δE_{NS} (‰)	Min., δE_{min} (‰)	Max., δE_{max} (‰)	Est., δE_{est} (‰)	Direct, QE_{NS}[1, 3] (kg/d)	Difference, QE_{dn}-QE_{up}[4] (kg/d)	NSQ.I[4] (kg/d)	NSX.I[4] (kg/d)
Dogwood Run (D1 to D2A): Septic (HW, BK2)													
C, particulate	881215	-26.97	6.66	-26.70	9.12	-23.03	-24.10	-21.90	-25.86	--	2.47	2.77	0.62
	890503	-26.45	--	--	--	-23.03	-24.10	-21.90	-	----	--	--	--
	890518	-26.90	244.29	-27.50	427.81	-23.03	-24.10	-21.90	-28.30	--	183.51	225.55	-66.24
	890706	-26.85	223.40	-26.90	271.31	-23.03	-24.10	-21.90	-27.13	--	47.91	56.46	-3.55
	891012	-27.50	8.08	-26.45	22.49	-23.03	-24.10	-21.90	-25.88	--	14.40	16.11	5.28
N, particulate	881215	2.33	.87	3.00	.93	.33	-2.10	3.00	12.72	--	.06	0	-.31
	890503	2.85	--	--	--	.33	-2.10	3.00	--	--	--	--	--
	890518	-1.10	28.19	.40	55.20	.33	-2.10	3.00	1.97	--	27.01	25.48	58.11
	890706	.20	21.53	1.85	30.15	.33	-2.10	3.00	5.98	--	8.61	64.35	397.92
	891012	.50	1.15	2.90	2.97	.33	-2.10	3.00	4.44	--	1.81	5.46	-40.67
N, dissolved	881215	2.20	.87	3.90	.93	16.29	1.40	31.19	28.55	--	.06	.10	.11
	890503	--	--	--	--	16.29	1.40	31.19	--	--	--	--	--
	890518	--	28.19	--	34.50	16.29	1.40	31.19	--	--	6.31	--	--
	890706	--	21.53	4.20	18.84	16.29	1.40	31.19	--	--	--	--	--
	891012	--	.87	--	2.22	16.29	1.40	31.19	--	--	1.36	--	--
S, particulate	881215	5.97	.98	5.50	1.68	3.18	1.00	4.15	4.91	--	.69	0	.28
	890503	5.20	--	--	--	3.18	1.00	4.15	--	--	--	--	--
	890518	2.60	202.01	6.10	324.31	3.18	1.00	4.15	11.88	--	122.29	-11.52	1,957.01
	890706	4.50	53.83	4.50	67.83	3.18	1.00	4.15	4.50	--	14.00	0	0
	891012	4.72	1.30	2.30	5.68	3.18	1.00	4.15	1.58	--	4.38	.54	8.92
S, dissolved	881215	3.60	.98	4.50	1.68	9.20	9.00	9.40	5.84	--	.69	.44	.27
	890503	--	--	--	--	9.20	9.00	9.40	--	--	--	--	--
	890518	4.90	187.92	--	324.31	9.20	9.00	9.40	--	--	136.39	--	--
	890706	4.60	53.83	6.10	67.83	9.20	9.00	9.40	11.87	--	14.00	18.06	22.12
	891012	1.70	1.30	7.00	4.94	9.20	9.00	9.40	8.89	--	3.64	3.52	3.49
Monroe Creek (M2A to M3): Tributary (M2B)													
C, particulate	880706	-24.30	26.31	--	--	--	-25.50	-25.50	--	--	--	--	--
	890510	-25.60	1119.02	-25.20	1341.12	--	-25.50	-25.50	-23.18	--	222.10	--	--
	890706	-26.10	691.09	-25.70	1098.83	-25.50	-25.50	-25.50	-25.02	103.33	407.74	400.10	732.55
N, particulate	880706	6.90	9.45	--	--	--	5.20	5.20	--	--	--	--	--
	890510	4.20	242.68	3.60	238.42	--	5.20	5.20	--	--	--	--	--
	890706	2.20	112.68	2.00	106.68	5.20	5.20	5.20	--	136.75	--	-19.07	-7.11
N, dissolved	880706	7.08	8.22	--	--	--	7.15	7.15	--	--	--	--	--
	890510	9.20	175.27	6.90	238.42	--	7.15	7.15	.52	--	63.15	--	--
	890706	5.80	67.61	6.60	106.68	7.15	7.15	7.15	7.98	136.75	39.08	43.63	63.22
S, particulate	880706	4.50	3.29	--	--	--	--	--	--	--	--	--	--
	890510	--	444.91	3.30	298.03	--	--	--	--	--	--	--	--
	890706	3.70	97.65	5.10	192.03	--	--	--	6.55	13.17	94.38	--	--
S, dissolved	880706	4.85	1.64	--	--	--	5.40	5.40	--	--	--	--	--
	890510	4.20	431.43	4.40	298.03	--	5.40	5.40	--	--	--	--	--
	890706	3.60	75.12	4.90	138.69	5.40	5.40	5.40	6.44	13.17	63.57	75.77	100.16

Table 8. Chemical loads and isotopic compositions of nitrogen sources estimated by mass balance in the Lower Susquehanna River Basin, Pennsylvania—Continued

[‰, per mil; kg/d, kilogram per day; mg/L, milligram per liter; ML/d, million liters per day; --, no data]

1	2	3	4	5	6	7	8	9	10	11	12	13	14
		Upstream[1]		Downstream[1]		N-source isotope composition[2]				Estimates of chemical load from N source			
												Isotope mass balance	
Surface-water constituent	Date sampled	Delta, δE_{up}[2] (‰)	Load, QE_{up}[3] (kg/d)	Delta, δE_{dn}[2] (‰)	Load, QE_{dn}[3] (kg/d)	Meas., δE_{NS} (‰)	Min., δE_{min} (‰)	Max., δE_{max} (‰)	Est., δE_{est} (‰)	Direct, QE_{NS}[1, 3] (kg/d)	Difference, QE_{dn}-QE_{up}[4] (kg/d)	NSQ.I[4] (kg/d)	NSX.I[4] (kg/d)
							Monroe Creek (M1 to M2A): Fertilizer (M2A)						
C, particulate	880706	-25.80	23.79	-24.30	26.31	-34.80	-42.50	-20.60	-10.14	--	2.52	.73	-4.38
	890510	-24.80	696.74	-25.60	1119.02	-34.80	-42.50	-20.60	-26.92	--	422.28	326.66	89.52
	890706	-26.30	616.68	-26.10	691.09	-34.80	-42.50	-20.60	-24.44	--	74.41	52.26	-16.26
N, particulate	880706	4.60	7.70	6.90	9.45	.67	-.90	3.10	16.92	--	1.76	12.76	-5.53
	890510	4.10	93.59	4.20	242.68	.67	-.90	3.10	4.26	--	149.09	424.69	-7.07
	890706	1.10	59.68	2.20	112.68	.67	-.90	3.10	3.44	--	53.00	126.85	-286.03
N, dissolved	880706	3.37	7.70	7.08	8.22	.67	-.90	3.10	62.02	--	.52	48.40	-11.28
	890510	8.70	93.59	9.20	175.27	.67	-.90	3.10	9.77	--	81.68	1197.32	-10.91
	890706	1.10	46.42	5.80	67.61	.67	-.90	3.10	16.10	--	21.19	511.59	-733.27
S, particulate	880706	4.90	2.80	4.50	3.29	9.00	-.50	19.90	2.21	--	.49	.82	-.32
	890510	-	207.98	-	444.91	9.00	-.50	19.90	--	--	236.93	--	--
	890706	3.60	86.20	3.70	97.65	9.00	-.50	19.90	4.45	--	11.45	9.26	1.81
S, dissolved	880706	2.70	2.80	4.85	1.64	9.00	-.50	19.90	--	--	--	.05	.56
	890510	4.30	207.98	4.20	431.43	9.00	-.50	19.90	4.11	--	223.45	101.96	-9.18
	890706	4.70	86.20	3.60	75.12	9.00	-.50	19.90	--	--	--	-14.97	-19.22
							Conestoga River, Field Site 1 (C1 WR to C1 WS): Manure (C1)						
C, particulate	880628	--	--	-22.70	66.35	-19.28	-20.60	-18.35	--	--	--	--	--
	881221	--	--	-25.60	241.26	-19.28	-20.60	-18.35	--	--	--	--	--
	890427	--	--	-23.40	315.13	-19.28	-20.60	-18.35	--	--	--	--	--
	890502	--	19.09	-23.90	485.96	-19.28	-20.60	-18.35	--	--	466.87	--	--
	890705	-19.80	--	-21.70	--	-19.28	-20.60	-18.35	--	--	--	--	--
N, particulate	880628	--	--	6.00	22.35	6.85	3.80	8.65	--	--	--	--	--
	881221	--	--	5.00	44.70	6.85	3.80	8.65	--	--	--	--	--
	890427	--	--	4.70	18.83	6.85	3.80	8.65	--	--	--	--	--
	890502	7.80	2.20	4.70	51.86	6.85	3.80	8.65	4.56	--	49.66	7.16	169.23
	890705	8.30	--	5.40	--	6.85	3.80	8.65	--	--	--	--	--
N, dissolved	880628	--	--	7.20	20.93	6.85	3.80	8.65	--	--	--	--	--
	881221	--	--	5.00	39.74	6.85	3.80	8.65	--	--	--	--	--
	890427	--	--	4.70	18.83	6.85	3.80	8.65	--	--	-	--	--
	890502	8.50	1.54	4.70	40.34	6.85	3.80	8.65	4.55	--	38.79	25.76	92.90
	890705	-.07	--	5.40	--	6.85	3.80	8.65	--	--	--	--	--
S, particulate	880628	--	--	4.20	33.35	4.14	3.40	5.00	--	--	--	--	--
	881221	--	--	4.70	37.61	4.14	3.40	5.00	--	--	--	--	--
	890427	--	--	5.30	12.88	4.14	3.40	5.00	--	--	--	--	--
	890502	2.40	2.72	3.65	34.57	4.14	3.40	5.00	3.76	--	31.86	0	24.79
	890705	3.75	--	4.20	--	4.14	3.40	5.00	--	--	--	--	--
S, dissolved	880628	--	--	5.10	33.35	4.14	3.40	5.00	--	--	--	--	--
	881221	--	--	5.00	30.51	4.14	3.40	5.00	--	--	--	--	--
	890427	--	--	4.90	12.88	4.14	3.40	5.00	--	--	--	--	--
	890502	2.40	2.72	4.60	34.57	4.14	3.40	5.00	4.79	--	31.86	36.81	43.63
	890705	3.75	--	3.20	--	4.14	3.40	5.00	--	--	--	--	-

[1]Loads of carbon, nitrogen, and sulfur transported within sampled reach of stream. In body of table, symbols in parentheses indicate locations of upstream, downstream, and source samples consistent with identification codes in figure 1 and Appendix tables A1, B2, B3, and B4.

[2]Delta, or δE, is isotopic composition, in per mil, of water and N-source samples, where E is ^{13}C, ^{15}N, or ^{34}S: δE_{up} is measured composition of upstream water particulate or dissolved fractions; δE_{dn} is measured composition of downstream water particulate or dissolved fractions; and δE_{NS} is measured composition of an additional source of the chemicals C, N, and S. The additional source, called the N source, can be a tributary stream, a point-source discharge, or a nonpoint-source contribution. Values for δE_{NS} either are concurrent with the upstream and downstream water samples or are an average of measured values for the N source. δE_{min} and δ_{max}, used for comparative purposes, are measured minimum and maximum isotope delta values for the N source. δE_{est} is the computed isotope delta value for the N source based on upstream and downstream loads (QE_{up}, QE_{dn}) and isotopic measurements (δE_{up}, δE_{dn}), where

$$\delta E_{est} = \frac{(\delta E \cdot QE)_{dn} - (\delta E \cdot QE)_{up}}{QE_{dn} - QE_{up}}$$

. This method corresponds with equation 5 in the text (mass-balance assumptions stated below in footnote 4).

[3]Load estimates by direct methods: QE_{up}, QE_{dn}, and QE_{NS} are for upstream, downstream, and chemical-source or tributary locations, respectively, which are computed as the product of the chemical concentration (mg/L) and discharge (ML/d) at each location.

[4]Load estimates by indirect, mass-balance methods: QE_{up}, QE_{dn}, NSQ.I, and NSX.I are for chemical-source or tributary addition to downstream load. Mass-balance estimates were computed using different equations, which follow:

1. $QE_{dn} - QE_{up}$: difference between downstream and upstream loads; does not require knowledge of isotopic composition.

2. $$NSQ.I = \frac{(\delta E \cdot QE)_{dn} - (\delta E \cdot QE)_{up}}{\delta E_{NS}}$$: isotope mass-balance difference between downstream and upstream loads. This method assumes knowledge of isotopic compositions of upstream, downstream, and N source and corresponds with a rearrangement of equation 5 in the text.

3. $$NSX.I = QE_{dn} \cdot \left(\frac{\delta E_{dn} - \delta E_{up}}{\delta E_{NS} - \delta E_{up}}\right)$$: isotope mass-balance difference. This method assumes knowledge of isotopic compositions of upstream water, downstream water, and N source, and corresponds with equation 4 in the text. Isotope-mass-balance methods assume (a) complete mixing of water at upstream, downstream, and chemical-source locations; (b) no isotopic fractionation; (c) isotopic composition of particulate fraction indicates the isotopic composition of the total load; and (d) isotopic composition of dissolved-inorganic fraction indicates the isotopic composition of the dissolved load.

(column 13) or by the isotopic difference ratio (column 14).

Point Sources

The effect of point-source contributions to the stream C, N, and S loads can be evaluated for water samples collected upstream and downstream from a sewage outflow pipe. On the basis of isotope mass balance, a simple mixing model should apply, and the downstream water should contain chemical loads and isotopic compositions that are the weighted averages of upstream water and sewage effluent. Table 8 shows computed loads, for each sampling date, in upstream (column 3) and downstream (column 5) water samples collected from Dogwood Run and in sewage effluent (columns 11–14) from the Dillsburg wastewater-treatment plant. The sewage chemical load computed from discharge and concentration (column 11) is within a factor of 2 but is not equal to estimates on the basis of downstream and upstream loads (column 12) and isotopic mass-balance computations (columns 13 and 14). For about one-half of the sampling dates, estimated values for N source $\delta^{13}C$, $\delta^{15}N$, and $\delta^{34}S$ (column 10) are within ±0.6 ‰ of the measured ranges for the sewage effluent (columns 8 and 9). However, several examples of negative and extremely dissimilar C, N, and S loads in columns 13 and 14 indicate an inconsistency in isotope delta values among the corresponding end-member water samples. In most of these examples of poor estimates of loads, the estimated isotope delta values also are far outside the measured range. These mixed results indicate potential errors from the assumption of constant isotopic compositions of the N source, represented by the mean, and from direct measurements of loads in stream-water samples as the product of concentration and discharge. A similar evaluation of point-source effects on isotopic compositions was conducted by measuring the load and isotopic compositions of dissolved and particulate C, N, and S in two merging streams and in the water downstream from the junction. Such an evaluation can be made for sampling points on Dogwood Run and on Monroe Creek (table 8). As was found with the sewage load estimates, the measured and estimated tributary loads are within a factor of 2 for only about one-half of the sampling dates for Dogwood Run and less than one-half for Monroe Creek. Generally, in these cases, the estimated isotopic compositions of the tributary are within ± 0.6‰ of the measured ranges of $\delta^{13}C$, $\delta^{15}N$, and $\delta^{34}S$ for the tributary samples. Extreme values, far outside the measured range, indicate the potential for errors in direct measurement of discharge and concentration.

In the point-source examples, the chemical loads from the particular N source or tributary are added to small, turbulent streams within short reaches, less than 100 m, between upstream and downstream sampling locations. No other sources are apparent in these reaches, and substantial inflow or outflow as ground-water seepage is unlikely. Differences in estimated loads shown in table 8, columns 11–14, are common, however. These differences are attributable to inconsistent data for variables in the computations. For example, isotope delta values in column 5 (downstream) should be in between values in columns 3 and 7 (upstream and N source, respectively). Errors in estimates of loads could result from inaccurate measurements of flow rates and concentrations of C, N, and S in the water samples. Errors also could result from assuming that the N source has a constant isotopic composition and from assuming that losses or gains of the chemicals and isotopic fractionation are not possible in the stream.

It is noteworthy that the estimated loads and isotopic compositions of the N sources on the basis of isotopic compositions of dissolved and particulate fractions in stream waters generally have about the same level of accuracy or error. Hence, the particulate fraction may be as useful a tracer of point sources as the dissolved fraction.

Nonpoint Sources

Loads from nonpoint sources including septic-field effluent (Dogwood Run), fertilizer (Monroe Creek), and land-spread manure (Little Conestoga Creek, Field Site 1) were estimated by use of the same approach that was used for point sources. Data to evaluate loads from each of these nonpoint sources are included in table 8. Loads estimated by various methods generally do not agree. The estimated values of $\delta^{15}N$ and $\delta^{34}S$ for the septic and manure sources are seldom within ±0.6‰ of the range of values measured for these materials (table 8). Estimated values for the fertilizer and associated stream waters at the Monroe Creek golf course generally differ from measured values, probably because the fertilizer is not a major source of the solutes in the stream. Nondetectable concentrations of N in some of the runoff-water and stream-water samples (table B3) indicate this effect. Hence, isotopic measurements undertaken for this study can only qualitatively confirm the potential water-quality effects by the presumed principal N sources.

Results of load estimates probably could be improved if samples of N-source materials (including rainfall), runoff waters, and stream waters were collected and measured concurrently. The first phase of the study emphasized the collection of N-source materials for characterization, and the second phase emphasized the collection of nearby soil and surface-water samples for qualitative comparison with N sources collected from the same locations, but commonly during different times.

On the basis of the results in table 8, there is little benefit from computing C, N, and S loads from isotopic compositions of nonpoint sources, mainly because (1) the source materials have widely variable isotopic compositions, (2) isotopic fractionation is likely to take place in the soils as the chemical compounds are processed during transport, and (3) anthropogenic N sources are not major C and S sources in agricultural soils. Relatively long residence times and slow transport rates through the soil will increase the potential for fractionating of the compounds. Long transport pathways and large reservoirs of the elements in the soils will also increase the potential for dilution of the N-source isotopic signature. Hence, unless the loading rate of anthropogenic compounds is large relative to reservoirs of the elements, or transport to receiving waters is relatively direct as with point sources, resolution for isotope tracing and mass balance for nonpoint sources will be limited. Qualitatively, however, the isotopic compositions of soil and particulate fractions appear to be useful as tracers of nonpoint sources because the local soil C, N, and S isotopic compositions are reflected in associated waters.

ESTIMATION OF N-ISOTOPIC FRACTIONATION

Although the concentration of N in agricultural soil is about one-tenth of that in forest-topsoil samples (tables 5 and B3), elevated concentrations of dissolved N in surface and ground waters from agricultural subbasins, especially manure-use areas, indicate an apparent imbalance in the local N cycle. This imbalance is reflected by lower C-org:N in the topsoil and subsoil of manure-use areas (respective medians of 11.7 and 10.4) and fertilizer-use areas (respective medians of 11.0 and 10.6) relative to those in forested areas (respective medians of 28.0 and 23.4). Processes leading to leaching of NO_3^- are likely to cause N isotopic fractionation, which can explain why values of $\delta^{15}N$ for agricultural soils and associated surface waters are more variable and distinctive than those of the forest system.

Soils and ground waters affected by animal manure typically contain large quantities of NO_3-N that is enriched in ^{15}N ($\delta^{15}N$ = +10‰ to +20‰). The hydrolysis of urea in animal manure and subsequent N-transformation reactions can produce this ^{15}N-enriched NH_4-N and NO_3-N. To explain the enrichment, the Rayleigh distillation equation and typical values of fractionation factors are used to estimate isotopic fractionation because of incomplete transformations during sequential steps of (1) ammonification, (2) volatilization, and (3) nitrification (fig. 12).

Although the assumptions are uncertain, figure 12 illustrates how a specific isotope delta value might result by assuming a series of reasonable reactions. Equation 9 may be utilized with kinetic fractionation factors, $\alpha_{s/p}$, of 1.007 for ammonification, 1.020 for volatilization, and 1.020 for nitrification (Letolle, 1980). Fresh steer manure and composted, liquefied dairy manure are 70 to 75 wt % organic N and 30 to 25 wt % NH_4^+. On the basis of these proportions, 30 percent of the organic N in manure, which has an initial $\delta^{15}N$ of 0.6‰ (reported for silage, Steele and Daniel, 1978) is assumed to be ammonified in the first step. The residual organic N will have $\delta^{15}N$ = +3.1‰, and product NH_4-N will have $\delta^{15}N$ = –5.2‰. In the second step, 50 percent of the NH_4-N is assumed to be lost by volatilization of NH_3-N (Denmeade and others, 1974), producing residual NH_4-N with $\delta^{15}N$ = +8.6‰ and NH_3 gas with –19.1‰. In the third step, 95 percent of the NH_4-N is nitrified (ratio of NO_3^-/NH_4^+ is about 20:1 in streams of "manure-use" areas), producing NO_3-N with $\delta^{15}N$ = +5.5‰. Thus, the final NO_3-N is enriched in ^{15}N relative to the initial organic material, and simply by varying the proportion of material reacted, the resultant isotopic compositions will vary.

The computed value of $\delta^{15}N$ = +8.6‰ for residual NH_4-N after the first step (fig. 12) is similar to median $\delta^{15}N$ values for N in manure, soil, and runoff-water particulate (table 6). The computed value of $\delta^{15}N$ for the final residual NH_4^+ after the last step is comparable with extremely large values of $\delta^{15}N$-NH_4 (from +30‰ to +42‰) that were measured in swine-manure lagoon effluent and associated surface water. The computed value of $\delta^{15}N$ = +5.5‰ for NO_3-N also is equivalent to the median $\delta^{15}N$ for dissolved nitrate in stream water but is somewhat greater than that of runoff and less than that of ground water. It is possible that denitrification in the subsurface caused the ground-water nitrate to become further enriched in ^{15}N.

Although the particulate fraction in runoff has a median $\delta^{15}N$ similar to that of manure and associated topsoil, dissolved nitrate in runoff has a low value of $\delta^{15}N$ = 2.3‰, which is substantially lower than that of manure and approaches values in a forested stream. This relatively low value is possible because of mixing with NO_3-N from rainfall, legumes, or fertilizer, which can have compositions of about $\delta^{15}N$ = 0‰. Thus, the combined effects of isotope fractionation and mixing of multiple sources can explain the variability in the data.

Figure 12 illustrates that measured N isotopic variations may be explained by considering transformation effects and reported values for fractionation factors. However, this very simple treatment of the data does not address all possible models for the evolution of N-isotopic compositions of soil and water in a specific land-use area. For example, other things being equal, but if 60 percent of the NH_4-N in manure is volatilized, the final NO_3-N would have $\delta^{15}N$ = 10.0‰, instead of 5.5‰.

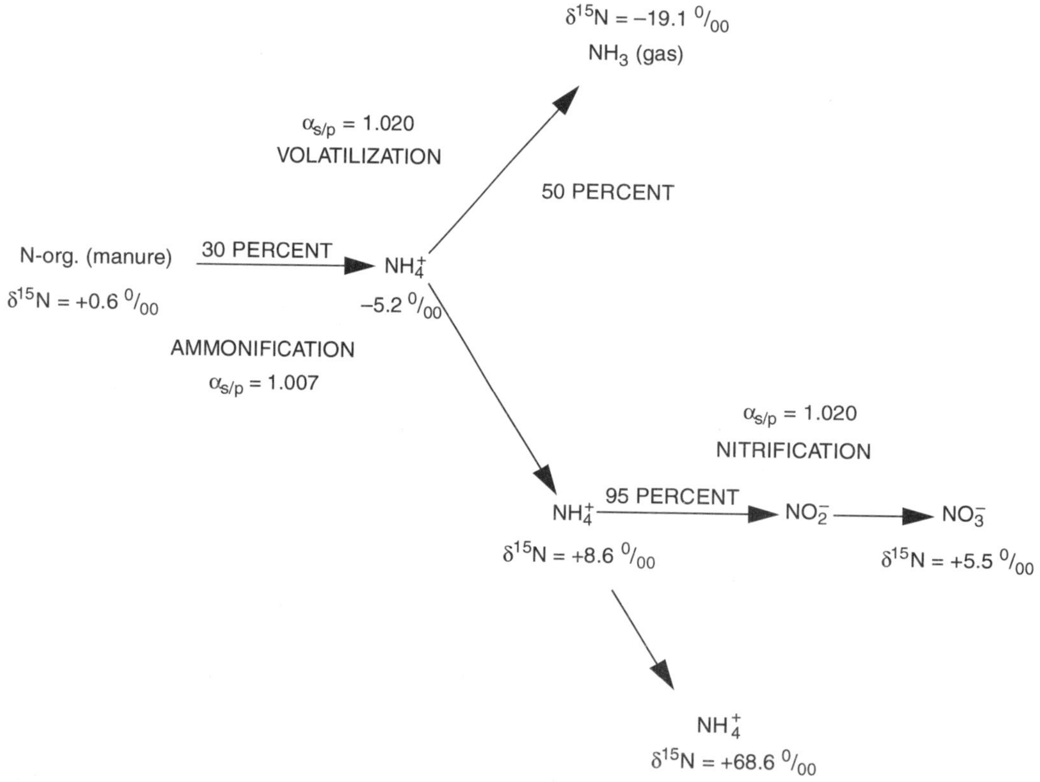

Figure 12. Nitrogen-isotopic fractionation by the series of reactions, ammonification, volatilization, and nitrification, in sequence.

PRACTICALITY OF USING STABLE ISOTOPES OF CARBON, NITROGEN, AND SULFUR TO IDENTIFY SOURCES OF NITROGEN IN SURFACE WATERS

One objective of this study was to obtain sufficient chemical and isotopic data to characterize the compositions of various N sources. These data provide helpful background information for studies of water-quality effects from different land-use activities. Although more than 150 samples were analyzed (table B1), because of the many different sample classes based on land use and sample medium, only about four samples, on average, were included in each class. Because the isotopic compositions vary widely within classes, greater numbers of samples would be helpful to account for spatial and temporal variations and to conduct more discriminating statistical analyses. However, regardless of the number of samples, several factors limit the general application of isotopic methods for tracing N pollution and quantifying contributions from different N sources in the watershed. These factors include temporal and spatial variability in biogeochemical processes, variability in loading rates affecting isotopic ratios, and difficulty in obtaining representative samples and accurate measurements of loading rates and isotopic compositions of rain-water and dilute surface-water samples.

Measurements

Problems in accurately measuring discharges and in analyzing dilute samples are apparent in the estimation of loads. Errors can be large relative to the quantities measured. Field measurements and sampling techniques can be refined. More accurate discharge measurements use calibrated weirs or flumes rather than wading measurements. However, for a reconnaissance-type investigation, such installations may be impractical. Isotopic measurement of dilute samples can be performed by use of techniques that concentrate the solutes of interest, such as by use of ion-exchange resins for rainfall studies (Hoering, 1957; Moore, 1974; Freyer, 1978). The distillation process to remove N from water samples for isotopic analysis is not ideal for low-concentration samples. Hence, it may be desirable to use chemical and isotopic analytical methods that are routinely used for rainfall on all the stream-water samples. However, such techniques are practiced by few laboratories.

Computations

Computations of mass balance are straightforward but are sensitive to imprecise values and propagation of errors. Computations of fractionation involve speculation regarding the extent of reaction progress and use of fractionation factors whose precise values depend on environmental and biological conditions. Limitations also are imposed by the fact that multiple processes and mixing of N sources are likely to cause isotopic variations. The combined effects of fractionation and mixing of sources can produce similar results, which do not have unique quantitative solutions. Therefore, limited isotopic measurements commonly provide only qualitative information, unless combined with other chemical and hydrologic data.

SUMMARY AND CONCLUSIONS

Samples of N-source material and associated nearby soil and water were collected from several small, primarily single-source, subbasins in the Lower Susquehanna River Basin, Pennsylvania, to determine whether stable isotopes of C, N, and S can be used to identify different N sources in stream waters. The data demonstrate that various N sources, including forest leaf litter, synthetic fertilizer, farm-animal manure, municipal-sewage effluent, and septic-tank effluent, and associated soils and waters have characteristic chemical and isotopic compositions.

The chemical-concentration data indicate that, with the exception of sewage and septic effluents, most N sources and soils contain larger proportions of organic and reduced forms of C, N, and S than inorganic, oxidized forms. In contrast, most surface water and ground water contain larger proportions of dissolved inorganic C, N, and S forms than organic forms. Furthermore, surface and ground water typically have C-org:N values that are much smaller than those of nearby, organic-rich N sources and soils. These data indicate that C, N, and S are extensively processed in soils and streams. The organic materials in soils and streams can be transformed into inorganic forms by respiration and oxidation, and inorganic forms can be converted into new organic compounds by photosynthesis, uptake, or assimilation. All these processes can cause isotopic fractionation, with a general tendency for the lighter isotopes to become concentrated in the products and the heavier isotopes to become concentrated in the residual reactant.

The isotopic data for the N sources indicate that animal manure, human waste (sewage plus septic), and forest leaf litter have distinctive $\delta^{13}C$ compositions. Most N sources do not have unique $\delta^{15}N$ and $\delta^{34}S$ compositions, however, owing to wide ranges of compositions within, and overlap among, different N-source types.

For isotopes to be useful as tracers of N sources, fractionation should be minimal during transport from the source to nearby surface waters, so the transported products will have isotope ratios similar to those of the source. In reality, however, fractionation does occur during transport. Consequently, the dissolved and particulate fractions of N and S in aqueous N-source and water samples were different from one another, although the average difference between $\delta^{34}S$ of dissolved and particulate fractions approached the precision of the overall method (±0.6‰). Furthermore, coexisting dissolved fractions of NO_3-N and NH_3-N in aqueous samples commonly had different isotopic compositions.

Although $\delta^{15}N$ values of soil and runoff-water samples are qualitatively similar to those of the applied N source, $\delta^{13}C$ and $\delta^{34}S$ for runoff-water and stream-water samples generally do not reflect those of the applied N source. Values of $\delta^{13}C$ for particulates and of $\delta^{34}S$ for particulates and dissolved SO_4^{2-} in the surface-water samples appear to reflect the compositions of soil organic matter and sulfur-bearing minerals, which likely are larger sources of the elements than the applied N-source material. Values of C-org:N combined with $\delta^{13}C$ aid in distinguishing agricultural soils (relatively high $\delta^{13}C$ and low C-org:N) from forested soils. The C-org:N values of suspended particulates in runoff or stream waters generally are lower than those of nearby soils, however, and indicate that oxidation of organic matter, other chemical transformations, and resultant isotopic fractionation can be important controls on the isotopic compositions of N-containing compounds in the soil and water.

Observed trends of lighter C and N isotopic compositions of forested topsoil relative to subsoil are consistent with other work (see Nadelhoffer and Fry, 1988). Similarly, lower values of $\delta^{13}C$ and $\delta^{15}N$ for the particulate fraction than for the dissolved fraction of forest stream water indicate that, even in a nitrogen-limited system, fractionation can be significant.

The relative uniformity of isotopic compositions, particularly $\delta^{34}S$, for stream waters in the study area was not helpful in identifying different sources of the elements or different land uses. However, this uniformity of compositions could be helpful for other applications. For example, streams sampled in this study contribute to C, N, and S loads transported from the Susquehanna River Basin to the Chesapeake Bay. The isotopic compositions of the stream samples could indicate terrigenous source contributions to food webs of the bay or other estuaries.

Isotopic measurements provide qualitative information about important reactions that can affect N concentrations in soils and surface waters. However, because of wide variations in source chemical and isotopic compositions and chemical transformations and fractionation during transport over short distances (hundreds of meters), mass-balance computations generally are not sufficiently accurate to estimate the proportions of multiple sources contributing to the N load in the streams studied. Uncertainties in mass-balance computations because of natural variations in compositions are complicated by errors associated with measurements of discharge, chemical concentrations, and isotopic compositions of relatively dilute, small streams.

Additional work to resolve the magnitude of effects from isotopic fractionation and from mixing of added nutrients with previously existing materials in soil and water would be helpful in the evaluation of the fate and transport of the nutrients and the computation of loads by mass balance. Knowledge of the concentrations and isotopic compositions of related organic and inorganic fractions in soil and water is critical for resolving effects of chemical transformations and fractionations. Instead of collecting data over a broad area, a local focus in a study area with specific, unchanging land use would be desirable. Detailed information on temporal and spatial variations in the C, N, and S compositions and loading rates to the local soil and water could be obtained. The use of sensitive and accurate methods for measuring discharge rates, low concentrations of chemical compounds and species, and corresponding isotopic compositions would minimize measurement errors and assure detection of variations in compositions and transport.

SELECTED REFERENCES

American Society for Testing and Materials, 1990, Annual book of ASTM standards: Philadelphia, Pa., ASTM, v. 11.01, p. 321–326.

Aravena, R., Evans, M.L., and Cherry, J.A., 1993, Stable isotopes of oxygen and nitrogen in source identification of nitrate from septic systems: Ground Water, v. 31, p. 180–186.

Baedecker, M.J., and Back, W., 1979, Hydrogeological processes and chemical reactions at a landfill: Ground Water, v. 17, p. 429–437.

Balesdent, J., Wagner, G.H., and Mariotti, A., 1988, Soil organic matter turnover in long-term field experiments as revealed by carbon-13 natural abundance: Soil Science Society of America Journal, v. 52, p. 118–124.

Bender, M.L., 1972, Carbon isotope fractionation, *in* Fairbridge, R.W., ed., Encyclopedia of Geochemistry and Environmental Sciences: New York, Van Nostrand Reinhold Co., p. 133–136.

Berg, B., and Staaf, H., 1981, Leaching, accumulation and release of nitrogen in decomposing forest litter, *in* Clark, F.E., and Rosswell, T., eds., Terrestrial nitrogen cycles—Processes, ecosystem strategies, and management impacts: Swedish Natural Science Research Council (NFR), Ecological Bulletins (Stockholm) 33, p. 163–178.

Bigeleisen, Jacob, 1965, Chemistry of isotopes: Science, v. 147, no. 3657, p. 463–471.

Black, C.A., Evans, D.D., White, J.L., Ensminger, L.E., and Clark, E.E., eds., 1965, Methods of soil analysis, Part 2. Chemical and microbiological properties: Madison, Wis., American Society of Agronomy Monograph No. 9.

Bolin, Bert, 1970, The carbon cycle: Scientific American, v. 223, p. 125–132.

Bolin, Bert, and Cook, R.B., eds., 1983, The major biogeochemical cycles and their interactions: New York, John Wiley & Sons, SCOPE 21, 451 p.

Bowen, Robert, 1988, Isotopes in the earth sciences: New York, Elsevier Applied Science Publishers Ltd., 647 p.

Boynton, W.R., Kemp, W.M., and Keefe, C.W., 1982, A comparative analysis of nutrients and other factors influencing estuarine phytoplankton production, *in* Kennedy, V.S., ed., Estuarine comparisons: New York, Academic Press, p. 69–90.

Brady, N.C., 1974, The nature and property of soils: New York, MacMillan Publishing Co., 639 p.

Bremner, J.M., and Tabatabai, M.A., 1973, Nitrogen-15 enrichment of soils and soil-derived nitrate: Journal of Environmental Quality, v. 2, p. 363–365.

Brezonik, P.L., 1973, Nitrogen sources and cycling in natural waters: U.S. Environmental Protection Agency EPA 660/3–73–002, 167 p.

Chambers, L.A., and Trudinger, P.A., 1979, Microbiological fractionation of stable sulfur isotopes—A review and critique: Geomicrobiology Journal, v. 1, p. 249–293.

Chapelle, F.H., and Knobel, L.L., 1985, Stable carbon isotopes of HCO_3^- in the Aquia aquifer, Maryland: Ground Water, v. 23, p. 592–599.

Chichester, D.C., 1988, Evaluation of agricultural best-management practices in the Conestoga River headwaters, Pennsylvania—Methods of data collection and analysis, and description of study areas: U.S. Geological Survey Open-File Report 88–96, 32 p.

Cifuentes, L.A., Fogel, M.L., Pennock, J.R., and Sharp, J.H., 1989, Biogeochemical factors that influence the stable nitrogen ratio of dissolved ammonium in the Delaware Estuary: Geochimica et Cosmochimica Acta, v. 53, p. 2713–2721.

Coplen, T.B., 1993, Uses of environmental isotopes, *in* Alley, W.M., ed., Regional ground-water quality: New York, Van Nostrand Reinhold, p. 227–254.

Coplen, T.B., Krouse, H.R., and Bohlke, J.K., 1992, Reporting of nitrogen-isotope abundances: Pure & Applied Chemistry, v. 64, p. 907–908.

Degens, E.T., 1969, Biogeochemistry of stable carbon isotopes, *in* Eglington, M.T., and Murphy, J., eds., Organic geochemistry: Berlin, Springer-Verlag, p. 304–329.

Deines, Peter, 1980, The isotopic composition of reduced organic carbon, *in* Fritz, P., and Fontes, J.Ch., eds., Handbook of environmental isotope geochemistry, v. 1. The terrestrial environment, A: Amsterdam, Elsevier, p. 329–406.

Delwiche, C.C., 1970, The nitrogen cycle: Scientific American, v. 223, p. 135–147.

Delwiche, C.C., and Steyn, P.L., 1970, Nitrogen isotope fractionation in soils and microbial reactions: Environmental Science & Technology, v. 4, no. 11, p. 929–935.

DeNiro, M.J., and Epstein, S., 1978, Influence of diet on the distribution of carbon isotopes in animals: Geochimica et Cosmochimica Acta, v. 42, p. 495–506.

———1981, Influence of diet on the distribution of nitrogen isotopes in animals: Geochimica et Cosmochimica Acta, v. 45, p. 341–351.

Denmeade, O.T., Nulsen, R., and Thurtell, G.W., 1978, Ammonia exchange over a corn crop: Soil Science Society of America Journal, v. 42, p. 840–842.

Denmeade, O.T., Simpson, J.R., and Freney, J.R., 1974, Ammonia flux into the atmosphere from a grazed pasture: Science, v. 185, p. 609–610.

Elliot, L.F., Schuman, G.E., and Viets, F.G., Jr., 1971, Volatilization of nitrogen-containing compounds from beef cattle areas: Soil Science Society of America Proceedings, v. 35, p. 752–755.

Exner, M.E., and Spalding, R.F., 1985, Ground-water contamination and well construction in southeast Nebraska: Ground Water, v. 23, p. 26–34.

Fairchild, J.F., Boyle, T.P., Robinson-Wilson, E., and Jones, J.R., 1983, Microbial action in detrital leaf processing and the effects of chemical perturbation, *in* Fontaine, T.D., III, and Bartell, S.M., eds., Dynamics of lotic ecosystems: Ann Arbor, Mich., Ann Arbor Science, p. 437–456.

Faure, Gunter, 1977, Principles of isotope geology: New York, John Wiley and Sons, 464 p.

Field, C.W., 1972, Sulfur—Element and geochemistry, *in* Fairbridge, R.W., ed., Encyclopedia of Geochemistry and Environmental Sciences: New York, Van Nostrand Reinhold Co., p. 1142–1148.

Fishel, D.K., Langland, M.J., and Truhlar, M.V., 1991, Hydrology and the hypothetical effects of reducing nutrient applications on water quality in the Bald Eagle Creek headwaters, southeastern Pennsylvania, prior to implementation of agricultural best-management practices: U.S. Geological Survey Water-Resources Investigations Report 91–4006, 59 p.

Fisher, D., Ceraso, J., Mathew, T., and Oppenheimer, M., 1988, Polluted coastal waters—The role of acid rain: New York, Environmental Defense Fund, 102 p.

Flipse, W.J., Jr., Katz, B.G., Lidner, J.B., and Markel, R., 1984, Sources of nitrate in ground water in a sewered housing development, central Long Island, New York: Ground Water, v. 22, p. 418–426.

Flipse, W.J., Jr., and Bonner, F.T., 1985, N-isotope ratios in ground water under fertilized fields, Long Island, New York: Ground Water, v. 23, p. 59–67.

Focht, D.D., 1973, Isotope fractionation of ^{15}N and ^{14}N in microbiological nitrogen transformations—A theoretical model: Journal of Environmental Quality, v. 2, no. 2, p. 247–252.

Freyer, H.D., 1978, Seasonal trends of NH_4^+ and NO_3^- nitrogen isotope composition in rain collected at Julich, Germany: Tellus, v. 30, p. 83–92.

Freyer, H.D., and Aly, A.I.M., 1974, Nitrogen-15 variations in fertilizer nitrogen: Journal of Environmental Quality, v. 3, no. 4, p. 405–406.

Fritz, P., and Fontes, J.Ch., 1980, Introduction, *in* Fritz, P., and Fontes, J.Ch., eds., Handbook of environmental isotope geochemistry, v. 1. The terrestrial environment, A: Amsterdam, Elsevier, p. 1–19.

Fry, Brian, 1986, Stable sulfur isotopic distributions and sulfate reduction in lake sediments of the Adirondack Mountains, New York: Biogeochemistry, v. 2, p. 329–343.

_____1988, Food web structure on Georges Bank from stable C, N, and S isotopic compositions: Limnology and Oceanography, v. 33, p. 1182–1190.

Fry, Brian, Brand, W., Mersch, F.J., Tholke, K., and Garritt, R., 1992, Automated analysis system for coupled $\delta^{13}C$ and $\delta^{15}N$ measurements: Analytical Chemistry, v. 64, p. 288–291.

Fry, Brian, Cox, J., Gest, H., and Hayes, J.M., 1986, Discrimination between ^{34}S and ^{32}S during bacterial metabolism of inorganic sulfur compounds: Journal of Bacteriology, v. 165, p. 328–330.

Fry, Brian, Scanlan, R.S., and Parker, P.L., 1977, Stable carbon isotope evidence for two sources of organic matter in coastal sediments—Seagrasses and plankton: Geochimica et Cosmochimica Acta, v. 41, p. 1875–1877.

Fry, Brian, and Sherr, E.B., 1984, $\delta^{13}C$ measurements as indicators of carbon flow in marine and freshwater ecosystems: Contributions to Marine Science, v. 27, p. 13–47.

Gillham, R.W., Cherry, J.A., Hendry, M.J., Sklash, M.G., Frind, E.O., Blackport, R.J., and Pucovsky, M.G., 1978, Studies of the agricultural contribution to nitrate and the subsequent loading to surface waters: Ontario, University of Waterloo, International Reference Group on Great Lakes Pollution from Land Use Activities, Agricultural Watershed Studies.

Goodell, H.G., 1972, Carbon/nitrogen ratio, *in* Fairbridge, R.W., ed., Encyclopedia of Geochemistry and Environmental Sciences: New York, Van Nostrand Reinhold Co., p. 136–142.

Gormly, J.R., and Spalding, R.F., 1979, Sources and concentrations of nitrate-N in ground water of the central Platte region, Nebraska: Ground Water, v. 17, p. 291–301.

Hauck, R.D., 1982, Nitrogen-isotope-ratio analysis, *in* Page, A.L., ed., Methods of soil analysis, Part 2 (revised): Madison, Wis., American Society of Agronomy, p. 735–780.

Hauck, R.D., and Tanji, K.K., 1982, N transfers and mass balances, *in* Stevenson, F.J., ed., Nitrogen in agricultural soils: Madison, Wis., American Society of Agronomy Monograph 22, p. 891–925.

Heaton, T.H.E., 1984, Sources of the nitrate in phreatic groundwater in the western Kalahari: Journal of Hydrology, v. 67, p. 249–259.

_____1986, Isotopic studies of nitrogen pollution in the hydrosphere and atmosphere—A review: Chemical Geology, v. 59, p. 87–102.

Heaton, T.H.E., Talma, A.S., and Vogel, J.C., 1983, Origin and history of nitrate in confined groundwater in the western Kalahari: Journal of Hydrology, v. 62, p. 243–262.

Helsel, D.R. 1987, Advantages of nonparametric procedures for analysis of water quality data: Hydrological Sciences Journal v. 32, no. 2, p. 179–190.

Hem, J.D., 1985, Study and interpretation of the chemical characteristics of natural water (3d): U.S. Geological Survey Water-Supply Paper 2254, 263 p.

Hoering, T.C., 1955, Variations of nitrogen-15 abundance in naturally occurring substances: Science, v. 122, p. 1233–1234.

Hoering, Thomas, 1957, The isotopic composition of the ammonia and the nitrate ion in rain: Geochimica et Cosmochimica Acta, v. 12, p. 97–102.

Horrigan, S.G., Montoya, J.P., Nevins, J.L., and McCarthy, J.J., 1990, Natural isotopic composition of dissolved inorganic nitrogen in the Chesapeake Bay: Estuarine, Coastal and Shelf Science, v. 30, p. 393–410.

Hubner, Heinrich, 1986, Isotope effects of nitrogen in the soil and biosphere, *in* Fritz, P., and Fontes, J.Ch., eds., Handbook of environmental isotope geochemistry, v. 2. The terrestrial environment, B: Amsterdam, Elsevier, p. 361–425.

James, W.F., Kennedy, R.H., Shain, W.E., and Myers, R.K., 1988, Leaf litter breakdown in a recently impounded reservoir: Water Resources Bulletin, v. 24, p. 831–837.

Jansson, S.L., and Persson, J., 1982, Mineralization and immobilization of soil N, *in* Stevenson, F.J., ed., Nitrogen in agricultural soils: Madison, Wis., American Society of Agronomy Monograph 22, p. 229-252.

Kaplan, I.R., 1972, Sulfur cycle, *in* Fairbridge, R.W., ed., Encyclopedia of Geochemistry and Environmental Sciences: New York, Van Nostrand Reinhold Co., p. 1148–1152.

_____1983, Stable isotopes of sulfur, nitrogen, and deuterium in recent marine environments, *in* Arthur, M.A., and others, eds., Stable isotopes in sedimentary geology: Dallas, Society of Economic Paleontologists and Mineralogists, v. 2, p. 1–108.

Kaplan, I.R., and Rittenberg, S.C., 1964, Microbiological fractionation of sulphur isotopes: Journal of General Microbiology, v. 34, p. 195–212.

Karamanos, R.E., and Rennie, D.A., 1980, Variations in natural nitrogen-15 abundance as an aid in tracing fertilizer nitrogen transformations: Soil Science Society of America Journal, v. 44, p. 57–62.

———1981, The isotope composition of residual fertilizer nitrogen in soil columns: Soil Science Society of America Journal, v. 45, p. 316–322.

Keeney, D.R., and Bremner, J.M., 1964, Effect of cultivation on the nitrogen distribution in soils: Soil Science Society of America Journal, v. 28, p. 653–656.

Komor, S.C., and Anderson, H.W., Jr., 1993, Nitrogen isotopes as indicators of nitrate sources in Minnesota sand-plain aquifers: Ground Water, v. 31, p. 260–270.

Kreitler, C.W., 1975, Determining the source of nitrate in ground water by N isotope studies: Austin, Texas, University of Texas Austin, Bureau of Economic Geology Report of Investigations No. 83, 57 p.

_____1979, Nitrogen-isotope ratio studies of soils and groundwater nitrate from alluvial fan aquifers in Texas: Journal of Hydrology, v. 42, p. 147–170.

Kreitler, C.W., and Browning, L.A., 1983, Nitrogen-isotope analysis of groundwater nitrate in carbonate aquifers—Natural sources versus human pollution: Journal of Hydrology, v. 61, p. 285–301.

Kreitler, C.W., and Jones, D.C., 1975, Natural soil nitrate—The cause of the nitrate contamination of ground water in Runnels County, Texas: Ground Water, v. 13, p. 53–61.

Kreitler, C.W., Ragone, S.E., and Katz, B.G., 1978, N^{15}/N^{14} ratios of ground-water nitrate, Long Island, New York: Ground Water, v. 16, p. 404–409.

Krouse, H.R., 1980, Sulphur isotopes in our environment, *in* Fritz, P., and Fontes, J.Ch., eds., Handbook of environmental isotope geochemistry, v. 1. The terrestrial environment, A: Amsterdam, Elsevier, p. 435–471.

Langland, Michael, 1992, Atmospheric deposition of ammonia from open manure-storage lagoons in south-central Pennsylvania: The Environmental Professional, v. 14, p. 28–37.

Legg, J.O., and Meisinger, J.J., 1982, Soil N budgets, *in* Stevenson, F.J., ed., Nitrogen in agricultural soils: Madison, Wis., American Society of Agronomy Monograph 22, p. 503–566.

Letolle, Rene, 1980, Nitrogen-15 in the natural environment, *in* Fritz, P., and Fontes, J.Ch., eds., Handbook of environmental isotope geochemistry, v. 1. The terrestrial environment, A: Amsterdam, Elsevier, p. 407–433.

Lindau, C.W., and Spalding, R.F., 1984, Major procedural discrepancies in soil extracted nitrate levels and N isotopic values: Ground Water, v. 22, p. 273–278.

Macko, S.A., Fogel, M.L., Hare, P.E., and Hoering, T.C., 1987, Isotopic fractionation of nitrogen and carbon in the synthesis of amino acids by microorganisms: Chemical Geology, v. 65, p. 79–92.

Mariotti, Andre, 1983, Atmospheric nitrogen is a reliable standard for natural ^{15}N abundance measurements: Nature, v. 303, p. 685–687.

Mariotti, Andre, Germon, J.C., Hubert, P., Kaiser, P., Letolle, R., Tardieux, A., and Tardieux, P., 1981, Experimental determination of nitrogen kinetic isotope fractionation—Some principles, illustration for the denitrification and nitrification processes: Plant and Soil, v. 62, p. 413–430.

Mariotti, Andre, Lancelot, Christiane, and Billen, Gilles, 1984, Natural isotopic composition of nitrogen as a tracer of origin for suspended organic matter in the Scheldt estuary: Geochimica et Cosmochimica Acta, v. 48, p. 549–555.

Mariotti, Andre, Landreau, Alain, and Simon, Beatrice, 1988, ^{15}N isotope biogeochemistry and natural denitrification process in groundwater—Application to the chalk aquifer of northern France: Geochimica et Cosmochimica Acta, v. 52, p. 1869–1878.

Mariotti, Andre, Pierre, D., Vedy, J.C., Bruckert, S., and Guillemot, J., 1980, The abundance of natural nitrogen-15 in the organic matter of soils along an altitudinal gradient: Catena, v. 7, p. 293–300.

Mendenhall, William, 1975, Introduction to probability and statistics (4th): North Scituate, Mass., Duxbury Press, 460 p.

Mook, W.G., Bommerson, J.C., and Staverman, W.H., 1974, Carbon isotope fractionation between dissolved bicarbonate and gaseous carbon dioxide: Earth and Planetary Science Letters, v. 22, p. 169-176.

Moore, Howard, 1974, Isotopic measurement of atmospheric nitrogen compounds: Tellus, v. 26B, p. 169-174.

_____1977, The isotopic composition of ammonia, nitrogen dioxide, and nitrate in the atmosphere: Atmospheric Environment, v. 11, p. 1239–1243.

Nadelhoffer, K.J., and Fry, B., 1988, Controls on natural nitrogen-15 and carbon-13 abundances in forest soil organic matter: Soil Science Society of America Journal, v. 52, p. 1633–1640.

Nelson, D.W., 1982, Gaseous losses of N other than through denitrification, *in* Stevenson, F.J., ed., Nitrogen in agricultural soils: Madison, Wis., American Society of Agronomy Monograph 22, p. 327-363.

Nissenbaum, A., and Schallinger, K.M., 1974, The distribution of the stable carbon isotope (13C/12C) in fractions of soil organic matter: Geoderma, v. 11, p. 137–145.

Noll, D.A., Bergstresser, T.W., and Woodcock, J., 1988, Overburden sampling and testing manual: Pennsylvania Department of Environmental Resources Bureau of Mining and Reclamation, 78 p.

Nommik, Hans, and Vahtras, Kaarel, 1982, Retention and fixation of ammonium and ammonia in soils, *in* Stevenson, F.J., ed., Nitrogen in agricultural soils: Madison, Wis., American Society of Agronomy Monograph 22, p. 123–172.

Nriagu, J.O., 1974, Fractionation of sulfur isotopes by sediment adsorption of sulfate: Earth and Planetary Sciences Letters, v. 22, p. 366–370.

Nriagu, J.O., and Coker, R.D., 1978, Isotopic composition of sulphur in atmospheric precipitation around Sudbury, Ontario: Nature, v. 274, p. 883–885.

———1983, Sulphur in sediments chronicles past changes in lake acidification: Nature, v. 303, p. 692–694.

Nriagu, J.O., and Glooschenko, W.A., 1992, Isotopic composition of sulfur in mosses across Canada: Environmental Science & Technology, v. 26, p. 85–89.

Nriagu, J.O., and Harvey, H.H., 1978, Isotopic variation as an index of sulphur pollution in lakes around Sudbury, Ontario: Nature, v. 273, p. 223–224.

O'Deen, W.A., and Porter, L.K., 1979, Digestion and collection of ammonia for nitrogen-15 and total nitrogen determination: Analytical Chemistry, v. 4, p. 586–589.

Officer, C.B., Biggs, R.B., Taft, J.L., Cronin, L.E., Tyler, M.A., and Boynton, W.R., 1984, Chesapeake Bay anoxia—origin, development, and significance: Science, v. 223, p. 22–27.

O'Leary, M.H., 1981, Carbon isotope fractionation in plants: Phytochemistry, v. 20, p. 553–567.

Olson, R.A., and Kurtz, L.T., 1982, Crop nitrogen requirements, utilization, and fertilization, *in* Stevenson, F.J., ed., Nitrogen in agricultural soils: Madison, Wis., American Society of Agronomy, Agronomy Monograph 22, p. 567–604.

Ott, A.N., Takita, C.S., Edwards, R.E., and Bollinger, S.W., 1991, Loads and yields of nutrients and suspended sediment transported in the Susquehanna River Basin, 1985–89: Susquehanna River Basin Commission Publication No. 136, 254 p.

Park, R., and Epstein, S., 1960, Carbon isotope fractionation during photosynthesis: Geochimica et Cosmochimica Acta, v. 21, p. 110–126.

Pearson, F.J., Jr., and Rightmire, C.T., 1980, Sulphur and oxygen isotopes in aqueous S compounds, *in* Fritz, P., and Fontes, J.Ch., eds., Handbook of environmental isotope geochemistry, v. 1., The terrestrial environment, Amsterdam: Elsevier Scientific Publishing Company, p. 229–258.

Peterson, B.J., and Fry, Brian, 1987, Stable isotopes in ecosystems studies: Annual Reviews of Ecological Systems, v. 18, p. 293–320.

Peterson, B.J., Howarth, R.W., and Garritt, R.H., 1985, Multiple stable isotopes used to trace the flow of organic matter in estuarine food webs: Science, v. 227, p. 1361–1363.

———1986, Sulfur and carbon isotopes as tracers of salt-marsh organic matter flow: Ecology, v. 67, p. 865–874.

P-STAT, 1989, P-STAT user's manual: Princeton, N.J., P-STAT, Inc., 2 vols.

Rast, Walter, and Sutton, J.E., 1989, Use of stable carbon and nitrogen isotopes to trace the larval striped bass food chain in the Sacramento-San Joaquin Estuary, California, April to September 1985: U.S. Geological Survey Water-Resources Investigations Report 88–4164, 62 p.

Rau, G.H., Sweeney, R.E., Kaplan, I.R., Mearns, A.J., and Young, D.R., 1981, Differences in animal ^{13}C, ^{15}N, and D abundance between a polluted and an unpolluted coastal site—Likely indicators of sewage uptake by a marine food web: Estuarine, Coastal and Shelf Science, v. 13, p. 701–707.

Reinhorn, Tova, and Avnimelech, Yoram, 1974, Nitrogen release associated with the decrease in soil organic matter in newly cultivated soils: Journal of Environmental Quality, v. 3, p. 118–121.

Rennie, D.A., and Paul, E.A., 1975, Nitrogen isotope ratios in surface and sub-surface soil horizons, *in* Isotope ratios as pollutant source and behavior indicators: Vienna, Austria, International Atomic Energy Agency, IAEA-SM-191, p. 441–454.

Rennie, D.A., Paul, E.A., and Johns, L.E., 1976, Natural nitrogen-15 abundance of soil and plant samples: Canadian Journal of Soil Science, v. 56, p. 43–50.

Ritter, W.F., and Chirnside, A.E.M., 1984, Impact of land use on ground-water quality in southern Delaware: Ground Water, v. 22, p. 38–47.

Rochow, E.G., 1977, Modern descriptive chemistry: Philadelphia, Pa., W.B. Saunders Co., 253 p.

Ryther, J.H., and Dunstan, W.M., 1971, Nitrogen, phosphorus, and eutrophication in the coastal marine environment: Science, v. 171, p. 1008–1013.

Sackett, W.M., 1989, Stable carbon isotope studies on organic matter in the marine environment, *in* Fritz, P., and Fontes, J.Ch., eds., Handbook of Environmental Isotope Geochemistry, v. 3, The marine environment, Amsterdam: Elsevier Scientific Publishing Company, p. 139–168.

SAS Institute, Inc., 1982a, SAS user's guide—Basics 1982 Edition: Cary, N.C., SAS Institute, Inc., 921 p.

———1982b, SAS user's guide—Statistics 1982 Edition: Cary, N.C., SAS Institute, Inc., 584 p.

———1988, Procedures guide, release 6.03 edition: Cary, N.C., SAS Institute Inc., 441 p.

Schwarcz, H.P., 1978, Isotopes in nature—Stable isotopes of carbon, *in* Wedepohl, K.H., ed., Handbook of geochemistry: Berlin, Springer-Verlag, v. II/1, chap. 6, p. B1–B15.

Shearer, G.B., Kohl, D.H., and Chien, S.H., 1978, The nitrogen-15 abundance in a wide variety of soils: Soil Science Society of America Journal, v. 42, p. 899–902.

Shearer, G.B., Kohl, D.H., and Commoner, B., 1974, The precision of determinations of the natural abundance of nitrogen-15 in soils, fertilizers, and shelf chemicals: Soil Science, v. 118, p. 308–316.

Smith, B.N., and Epstein, S., 1971, Two categories of $^{13}C/^{12}C$ ratios for higher plants: Plant Physiology, v. 47, p. 380–384.

Spalding, R.F., Exner, M.E., Lindau, C.W., and Eaton, D.W., 1982, Investigation of sources of groundwater nitrate contamination in the Burbank-Wallula area of Washington, U.S.A.: Journal of Hydrology, v. 58, p. 307–324.

Spalding, R.F., and Fulton, J.W., 1988, Groundwater munition residues and nitrate near Grand Island, Nebraska, U.S.A.: Journal of Contaminant Hydrology, v. 2, p. 139–153.

Spiker, E.C., and Kendall, Carol, 1983, C-13 and N-15 as source indicators of sedimentary organic matter in the Chesapeake Bay system [abs.]: Estuaries, v. 6, p. 305.

Sprent, J.I., 1987, The ecology of the nitrogen cycle: Great Britain, Cambridge University Press, Cambridge Studies in Ecology, 151 p.

Stam, A.C., Mitchell, M.J., Krouse, H.R., and Kahl, J.S., 1992, Stable sulfur isotopes of sulfate in precipitation and stream solutions in a northern hardwood watershed: Water Resources Research, v. 28, p. 231–236.

Steele, K.W., and Daniel, R.M., 1978, Fractionation of nitrogen isotopes by animals—A further complication to the use of variations in the natural abundance of ^{15}N for tracer studies: Journal of Agricultural Science, v. 90, p. 7–9.

Sternberg, Leonel, and DeNiro, M.J., 1983, Isotopic composition of cellulose from C_3, C_4, and CAM plants growing near one another: Science, v. 220, p. 947–949.

Stevenson, F.J., 1962, Chemical state of the nitrogen in rocks: Geochimica et Cosmochimica Acta, v. 26, p. 797–809.

———1972a, Nitrogen—Element and geochemistry, *in* Fairbridge, R.W., ed., Encyclopedia of Geochemistry and Environmental Sciences: New York, Van Nostrand Reinhold Co., p. 795–801.

———1972b, Nitrogen cycle, *in* Fairbridge, R.W., ed., Encyclopedia of Geochemistry and Environmental Sciences: New York, Van Nostrand Reinhold Co., p. 801–806.

———1982, Origin and distribution of N in soil, *in* Stevenson, F.J., ed., Nitrogen in agricultural soils: American Society of Agronomy Monograph 22, p. 1–42.

Stumm, Werner, and Morgan, J.J., 1981, Aquatic chemistry—An introduction emphasizing chemical equilibria in natural waters (2d): New York, Wiley-Interscience, 780 p.

Teeri, J.A., and Schoeller, D.A., 1979, $\delta^{13}C$ values of an herbivore and the ratio of C_3 and C_4 plant carbon in its diet: Oecologia, v. 39, p. 197–200.

Teply, L.J., Alberty, R.A., Carter, H.E., Kucera, T.J., Nixon, A.C., Phillips, D.L., and Wilson, L.S., 1980, Chemistry and the food system—A study by the Committee on Chemistry and Public Affairs: Washington, D.C., American Chemical Society, 138 p.

Thode, H.G., 1972, Sulfur isotope fractionation in biological processes, *in* Fairbridge, R.W., ed., Encyclopedia of Geochemistry and Environmental Sciences: New York, Van Nostrand Reinhold Co., p. 1152–1157.

Thode, H.G., Monster, J., and Dunford, H.B., 1961, Sulphur isotope geochemistry: Geochimica et Cosmochimica Acta, v. 25, p. 150–174.

Tholander, Bent, 1977, An example of design of activated sludge plants with denitrification: Progress in Water Technology, v. 8, nos. 4/5, p. 661–672.

Tiessen, H., Karamanos, R.E., Stewart, J.W.B., and Selles, F., 1984, Natural nitrogen-15 abundance as an indicator of soil organic matter transformations in native and cultivated soils: Soil Science Society of America Journal, v. 48, p. 312–315.

Toran, Laura, 1982, Isotopes in ground-water investigations: Ground Water, v. 20, no. 6, p. 740–745.

U.S. Department of Agriculture, 1992, Conestoga headwaters project, Pennsylvania, Rural clean water program 10-year report, 1981–1991: U.S. Department of Agriculture, Agricultural Stabilization and Conservation Service, 329 p. (U.S. GPO: 1992–624–121:60346).

U.S. Environmental Protection Agency, 1979, Methods for chemical analysis of water and wastes: Cincinnati, Ohio, U.S. EPA, n.p.

Velinsky, D.J., Pennock, J.R., Sharp, J.H., Cifuentes, L.A., and Fogel, M.L., 1989, Determination of the isotopic composition of ammonium-nitrogen at the natural abundance level from estuarine waters: Marine Chemistry, v. 26, p. 351–361.

Velleman, P.F., and Hoaglin, D., 1981, Applications, basics, and computing of exploratory data analysis: Boston, Mass., Duxbury Press, 354 p.

White, J.R., Gubala, C.P., Fry, B., Owen, J., and Mitchell, M.J., 1989, Sediment biogeochemistry of iron and sulfur in an acidic lake: Geochimica et Cosmochimica Acta, v. 53, p. 2547–2559.

Wolterink, T.J., Williamson, H.J., Jones, D.C., Grimshaw, T.W., and Holland, W.F., 1979, Identifying sources of subsurface nitrate pollution with stable nitrogen isotopes: U.S. Environmental Protection Agency, EPA–600/4–79–050, 150 p.

Yanagisawa, Fumitaka, and Sakai, Hitoshi, 1983, Thermal decomposition of barium sulfate-vanadium pentaoxide-silica glass mixtures for preparation of sulfur dioxide in sulfur isotope ratio measurements: Analytical Chemistry, v. 55, p. 985–987.

APPENDIX A:

Descriptions of subbasins and sample sites in the Lower Susquehanna River Basin, Pennsylvania

Figure 1 shows the locations of the sample sites, and table A1 summarizes the land use and predominant N sources at these sites in the subbasins described in Appendix A.

STONY CREEK

An undeveloped area in the Stony Creek Basin, an elongate steep-sided valley in Dauphin County, was selected to study the background composition of stream water and soils from natural forest lands. The 5,700-ha watershed area upstream from the sampling location (SC in fig. 1) is underlain primarily by sandstone. Thin sandy loam soils have developed beneath an organic-rich topsoil horizon, which consists of several inches of decaying leaf litter and wood. Discharge rates ranging from about 310 to 2,000 L/s were measured intermittently during the study period (table B2) by wading 10 m across Stony Creek. Relatively low concentrations of suspended sediment (less than 1.0 mg/L), low specific-conductance values (S.C. = 22–27 µS/cm), variable pH (5.8–8.4), and relatively constant temperature (11–12° C) (table B2) are typical of a pristine, forested watershed in noncarbonate-rock terranes, where runoff is negligible.

MONROE CREEK

A 19-ha golf course in Lebanon County, along the lower reaches of Monroe Creek (M1–M4) (fig. 1), was selected to study the potential effects of synthetic fertilizer used in turf management on the N load in a small stream. Monroe Creek is a perennial stream that has its headwaters in a forested area, passes through sparsely developed rural-agricultural lands, and then flows unobstructed about 2 km through the golf course, which uses only solid, N-rich synthetic fertilizer mixtures. The 1,800-ha watershed area upstream from sampling locations at the golf course consists of a steep-sided valley underlain by shale and sandstone and is about 75 percent forested. The soil at the golf course is a thin, clay loam. Land uses upstream and surrounding the golf course are low-density rural residential and light agricultural; large areas of the valley are covered by alfalfa, hay, and some corn fields. A series of three stream-water sample sites on Monroe Creek are located upstream (M1), midstream (M2A), and downstream (M3) of the golf course (table B2). A mixed-source tributary below the midstream sample point (M2B), a spring in the vicinity of the downstream sample point (M2C), plus rainfall, runoff, and soil at a green and adjacent fairway (M2A) also were selected for sampling. Discharge rates ranging from about 48 to 2,000 L/s were measured intermittently during the study period (table B2) by wading 5 m across Monroe Creek at site M2A, where relatively low concentrations of suspended sediments (9–18 mg/L), low specific-conductance values (57–67 µS/cm), constant pH (6.8–7.1), and constant temperature (11–14° C) (table B2) were measured. The relatively unchanged chemical measurements during low and high streamflow conditions indicate that the runoff component is minor. These measurements are characteristic of a largely forested, and only lightly developed, watershed in noncarbonate-rock terranes.

BALD EAGLE CREEK

A 14-ha farm in York County, at the headwaters of Bald Eagle Creek (BE in fig. 1) (Fishel and others, 1991), was selected to study the potential effects of synthetic fertilizer used in agriculture (for growing corn, wheat, soybeans, and potatoes) on the N load in a small stream. The creek at the field site, where only synthetic fertilizer is applied on the land, normally is dry but, following intense or prolonged rainfall or snowmelt, will flow for periods of days to weeks. Downstream from the field site, Bald Eagle Creek is perennial and flows through pastures where dairy and beef cattle graze. A U.S. Geological Survey weir located about 3 km downstream from the headwaters field site was used to measure discharge from the 111-ha, mixed-source (fertilizer plus animal manure) watershed (Fishel and others, 1991). The watershed area above the weir is underlain by a quartz schist and is 100 percent cultivated. Soil is a sandy, micaceous loam, which erodes readily. Sites selected for runoff-water and soil sampling (BE1) are located in the headwaters field, and a stream-water site (BE2) is located at the U.S. Geological Survey weir downstream. During the study period, intermittently measured discharge rates at site BE2 ranged widely from about 3 to 140 L/s. Water-quality characteristics varied accordingly. Concentrations of suspended sediments ranged from less than 1 to 850 mg/L, specific conductance from 100 to 186 µS/cm, pH from 6.6 to 7.8, and temperature from 3° to 15° C (table B2). The variability of these measurements indicates that runoff constitutes an important component in high-streamflow conditions and is characteristic of a cultivated (tilled), sloping area in a noncarbonate-rock watershed.

BRUSH RUN

A 98-ha farm in Adams County, at the headwaters of Brush Run (BR in fig. 1) (Langland, 1992), was selected to study the potential effects of profuse manure spreading on the N load in a small stream. The stream channel extends about 2 km across the farm, and streamflow becomes perennial along this reach. During low-flow periods, the middle reaches of Brush Run through the farm (BR1 and BR2 in table B2) consisted of a series of pools separated by sections of exposed streambed. The 109-ha watershed area for Brush Run inclusive of the farm is underlain by shale and is covered by about 90 percent cultivated land with the remainder by farm structures. The soil is a micaceous, silty clay. Crops include wheat, corn, and soybeans, much of which is used onsite for animal feed. Swine and chickens are raised in stock houses, and the animal manure is spread on the adjacent cultivated fields, which are underlain by tile drains discharging to Brush Run (Langland, 1992). Liquid swine manure is stored in an unlined lagoon at the Brush Run streambank (BR1MS) and is sprayed on adjacent fields by use of an automatic sprinkler system. Soil sampling locations (BR1S) are in these fields. Stream-water sample sites on Brush Run are located at a U.S. Geological Survey weir (BR1) adjacent to the manure lagoon and about 50 m downstream near a pond (BR2). Discharge rates of less than 1 to 160 L/s were measured intermittently at the weir at site BR1 during the study period (table B2). At site BR2, downflow from the manure lagoon, variable concentrations of suspended sediments (8–54 mg/L), extremely variable, large values of specific conductance (300–8,100 µS/cm), relatively constant pH (7.4–7.9), and variable temperature (less than 1° –32° C) (table B2) were measured. The high specific conductance (8,100 µS/cm) and pH (7.9) of the water in Brush Run indicate probable contamination from the swine manure (S.C. 11,500 µS/cm and pH 7.8) (table B2).

CONESTOGA RIVER

Two farm-field "runoff" sites, Field-Site 1 (C1) and Field-Site 2 (C2), which have been monitored by the U.S. Geological Survey as part of a long-term study of the effects of manure management and intensive agriculture on ground-water quality in the Conestoga River headwaters in Lancaster County (Chichester, 1988; U.S. Department of Agriculture, 1992) (fig. 1) were selected for study. Each field site consists of more than 90 percent cultivated land and has clay-loam soils underlain by limestone and dolomite.

Field-Site 1

Field-Site 1 (C1) is a 9.3-ha dairy farm, which consists of several barns and cultivated, terraced fields of corn, alfalfa, and wheat. The main barn, where cows are raised and milked, is connected to a cement-lined manure-storage pit by a conveyor trench. The manure is spread periodically on the adjacent field above the main barn. An unnamed, ungaged tributary stream to the Conestoga River flows through the pasture at the farm about 50 m below the barn and manure pit. Periodic runoff-water (C1WR) and soil-sample sites (C1S) were located in the field. A site on the stream (C1WS), downstream from the entry of runoff and a spring discharge (SP58), was selected for collecting instantaneous stream-discharge measurements and water-quality samples. An additional dairy farm (CLM) located near the headwaters of Little Conestoga Creek also was selected for the collection of dairy manure. Field measurements of relatively constant near-neutral to alkaline pH (7.1–7.8) and constant, relatively high specific conductance (390–400 µS/cm), and variable temperature of the stream (4° –17.5° C) are characteristic of calcium-bicarbonate surface waters draining areas underlain by carbonate rocks. The runoff water has a lower specific conductance (160 µS/cm), however, because mineralized ground waters are not the major source of solutes in runoff.

Field-Site 2

Field-Site 2 consists of a 19-ha farm along the bank of Indian Creek, which is a tributary to the Conestoga River. Crops include corn and soybeans, much of which are used onsite for animal feed, plus tobacco and tomatoes. Steers, swine, and chickens are raised in stock houses for meat production. The steer and chicken manure is cleaned out of the animal-housing structures and spread directly on the adjacent fields upslope from the structures. Liquid swine manure is stored in a cement-lined pit and is injected into the soil of the fields several times each year. Periodic runoff-water (C2AWR, C2BWR) and soil-sample sites (C2S) were located in the fields. In addition, ground water, which discharges from a diffuse-flow spring (C2WG, referred to in previous studies as SP 61) below the fields and barns and flows directly to Indian Creek, is accessible for sampling. Field measurements of relatively constant near-neutral pH (6.8–7.5) and constant, high specific conductance (720–740 µS/cm) of the spring water are characteristic of carbonate ground waters. The runoff water has a much more variable specific conductance (200–2,200 µS/cm), however, because of resolubilization of solutes from liquid manure spread on the field.

CODORUS CREEK

Two stream-gaging stations on Codorus Creek in York County, near York, were selected to study the effects of effluent from urban sewage treatment (wastewater) on the N load in a stream draining mixed land uses. Water-sample sites were located at York (CCY) and at Pleasureville (CCP), upstream and downstream, respectively, from the York wastewater-treatment plant (YW). During the low-flow sampling event of June 1988 (table B2), discharge rates at U.S. Geological Survey gages at York and at Pleasureville were 1,700 and 2,950 L/s, respectively. Although several small streams flow into the creek along this reach, the increase of 1,250 L/s from upstream to downstream sampling points during the base-flow condition can be accounted for by the average discharge from the York wastewater plant (1,300 L/s). The wastewater effluent has higher specific conductance (1,020 and 1,100 µS/cm) and temperature (24.5° C) than those measured for upstream and downstream samples (specific conductance = 850 and 890 µS/cm and temperature = 21.0° and 21.5° C, respectively) (table B2). The increased specific conductance and temperature in Codorus Creek below the sewage plant indicate possible influence from the sewage plant.

DOGWOOD RUN

The Dogwood Run Basin, in Cumberland County, was selected to study the effects of single-source and mixed-source influences on water quality. The 2,300-ha watershed area is underlain by sandstone colluvium in the forested headwaters and by limestone in developed areas downstream to the Yellow Breeches Creek. The subsoil varies accordingly from a rocky, sandy loam in the headwaters to a clay loam downstream. Topsoil in the forested area consists of 5–10 cm of leaf litter similar to that found in the Stony Creek Basin; elsewhere, cultivated pasture and lawns prevail. Land uses change downstream—forested conditions (D1) to rural development, with septic fields (D2A), to urban and light industrial development (D2B) in Dillsburg with sewage-treatment effluent (DW) being discharged to Dogwood Run (D4) below Dillsburg. Field water-quality measurements (table B2) indicate that specific conductance increases progressively downstream as discharge increases, which is expected as ground water from carbonate-rock aquifers contributes to the streamflow and as septic and agricultural land uses become prevalent in the watershed.

In the lower part of the Dogwood Run Basin, an unnamed tributary stream flows through Berkshire Hills residential development, which contains a high density of septic fields. Stream-water sampling sites were located upstream (BH1) and downstream (BH2) from the development. Field data, however, indicate essentially unchanged flow rates and water quality between these two locations, which indicates that septic influence is unlikely to be observable.

SEWAGE-TREATMENT PLANTS

Most wastewater from domestic, food-processing, and industrial uses in urban and suburban areas of the lower Susquehanna River Basin is processed at municipal treatment plants located along tributaries and the main stem of the Susquehanna River. The Dillsburg, York, and Harrisburg wastewater-treatment plants were selected for study of point-source, treated sewage effluent. The Dillsburg sewage-treatment plant is a small facility, which discharges on average about 180 L/s. Sewage from Dillsburg and vicinity is entirely domestic waste. The influent is mechanically pulverized, treated with alum to remove phosphorus, aerated with compressed air, and then decomposed in two activated sludge pools. The settleable sludge, which is mostly bacteria, is recycled, and the supernatant (effluent) is chlorinated and then discharged into Dogwood Run.

The York wastewater-treatment plant is much larger than the Dillsburg facility. The York plant processes about 1,300 L/s of wastewater from York and vicinity. About 60 percent of the influent and 40 percent of the biochemical oxygen demand (BOD) are from food processing and industrial facilities. The remainder is domestic sewage. The influent is treated with polymers to assist in settling primary sludge, which is about 2 to 4 percent solids. The primary sludge is digested anaerobically, which creates methane, and then is pressed into filter cake, which is about 90 percent water. The filter-cake sludge is considered hazardous waste and is landfilled because of the heavy-metal content. Until about October 1988, the supernatant was treated either by contact stabilization (9,300 L/s) or oxygenated aeration (6,200 L/s), and then chlorinated prior to discharge. The contact stabilization process was replaced in October 1988 by a 13,400-L/s anaerobic-oxic (A-O) process that uses compressed air and biological removal of N and P through a succession of anaerobic to aerobic steps. The modernized facility combines effluent from the pure-oxygen treatment method and the A-O process. On average, about 1,300 L/s of effluent is discharged into Codorus Creek.

The Harrisburg wastewater-treatment plant is about the same size as the York plant, but the Harrisburg plant influent is about 90 percent domestic sewage and uses oxygenated aeration along with anaerobic digestion. In addition to sewage received by pipeline, the plant also receives and treats three to five pump-truck loads (30,000–50,000 L) of domestic septic-tank waste daily and also primary sludge from smaller, less advanced treatment plants. The trucked-in sludge is mixed with the sewer-line influent. The combined influent is separated into sludge and supernatant and then processed. The sludge filter cake is normally incinerated but also is landfilled. The effluent is aerated with oxygen, neutralized, and chlorinated before being discharged at a rate of about 1,100 L/s into the Susquehanna River.

Table A1. Locations of sample-collection sites in the Lower Susquehanna River Basin, Pennsylvania

[lat-long-no, latitude, longitude, and site identifier number; n.a., not applicable; km^2, square kilometer]

Site name	Site number (lat-long-no)	Local ID	County	Principal upstream land use	Principal lithology	Upstream area (km^2)
Bald Eagle Creek (runoff)	394504-762851-01	BE1	York	Fertilizer, agriculture	Schist	0.14
Bald Eagle Creek (headwaters)	394504-762851-01	BE1	York	Fertilizer, agriculture	Schist	.14
Bald Eagle Creek (at gage)	394454-762750-01	BE2	York	Mixed, fertilizer + manure	Schist	1.11
Berks well water	403051-761115-01	BK1	Berks	Forested	Shale	n.a.
Berks septic effluent	403050-761114-02	BK2	Berks	Septic, single house	Shale	n.a.
Berkshire Hills (upstream)	400824-770139-01	BH1	Cumberland	Rural undeveloped	Dolomite	.26
Berkshire Hills (downstream)	400836-770140-02	BH2	Cumberland	Septic, housing subdivision	Dolomite	.30
Brush Run (upstream, weir)	394906-770626-02	BR1	Adams	Manure, swine + chicken	Shale	.98
Brush Run (downstream, pond)	394911-770625-03	BR2	Adams	Manure, swine + chicken	Shale	1.09
Brush Run swine manure	394906-770626-01	BR1	Adams	Manure, swine + chicken	Shale	n.a.
Codorus Creek, York	395646-764520-01	CCY	York	Mixed, locally developed	Shale	575
Codorus Creek, Pleasureville	400107-764136-01	CCP	York	Sewage urban	Shale	692
York Wastewater (oxygen)	395917-764327-01	YWO	York	Sewage urban	Shale	n.a.
York Wastewater (contact)	395914-764327-02	YWD	York	Sewage urban	Shale	n.a.
Dogwood Run #1 (upstream)	400614-770512-01	D1	York	Forested	Colluvium	11.1
Dogwood Run #2A (midstream)	400653-770230-01	D2A	York	Septic, rural lots	Colluvium	16.1
Dogwood Run #2B (tributary)	400653-770229-01	D2B	York	Mixed, suburban runoff	Limestone	5.7
Dogwood Run #3 (pre-sewage)	400659-770232-01	D3	York	Mixed, septic + suburban	Limestone	22.5
Dogwood Run #4 (post-sewage)	400702-770233-01	D4	York	Sewage suburban	Limestone	22.8
Dillsburg Wastewater	400700-770232-01	DWA	York	Sewage suburban	Limestone	n.a.
Monroe Creek #1 (upstream)	402831-762839-01	M1	Lebanon	Mixed, mostly undeveloped	Shale	18.1
Monroe Creek #2A (midstream)	402810-762944-01	M2A	Lebanon	Fertilizer, golf course	Shale	18.6
Monroe Creek #2B (tributary)	402811-762958-01	M2B	Lebanon	Mixed, fertilizer + manure?	Shale	2.85
Monroe Creek #3 (downstream)	402810-763003-01	M3	Lebanon	Mixed, fertilizer + septic?	Shale	21.5
Monroe Creek #2A (runoff)	402810-762944-02	M2A	Lebanon	Fertilizer, golf course	Shale	.08
Monroe Creek #2C (spring)	402811-763003-01	M2C	Lebanon	Mixed, septic?	Shale	n.a.
Conestoga FS#1 (stream)	400745-755837-01	C1	Lancaster	Mixed, manure + septic	Limestone	15
Conestoga FS#1 (runoff)	400742-755840-02	C1	Lancaster	Manure, dairy	Limestone	.09
Conestoga LM	400847-755537-03	CLM	Lancaster	Manure, dairy	Limestone	3.7
Conestoga FS#2 (spring)	401150-761053-01	C2	Lancaster	Manure, swine + chicken + steer	Limestone	n.a.
Conestoga FS#2 (runoff)	401150-761053-01	C2	Lancaster	Manure, swine + chicken + steer	Limestone	.19
Stony Creek (near gate)	402440-764739-01	SC	Dauphin	Forested	Sandstone	57.0
Harrisburg Wastewater (oxygen)	401419-765120-01	HWO	Dauphin	Sewage urban	n.a.	n.a.
Harrisburg septic-tank	401419-765120-02	HW	Dauphin	Septic tank	n.a.	n.a.

APPENDIX B:

Field, chemical, and isotopic data for nitrogen-source and associated soil and water samples from the Lower Susquehanna River Basin, Pennsylvania, 1988–90

Table B1. Summary of types and dates of collection of samples from the Lower Susquehanna River Basin, Pennsylvania

Sample type	Number collected during specific period and conditions					Total collected in study
	6/88–7/88 Low flow	12/88–1/89 Low flow	4/89–5/89 High flow	6/89–7/89 High flow	10/89–1/90 Low flow	
Synthetic fertilizer	3	0	3	0	0	6
Liquid N solution	1	0	0	0	0	
Solid N-P-K mixture	2	0	3	0	0	
Animal manure	6	7	5	0	0	18
Dairy cattle (cow)	1	2	1	0	0	
Feeder cattle (steer)	1	1	1	0	0	
Swine	2	3	2	0	0	
Chicken	2	1	1	0	0	
Human waste	7	3	7	1	3	21
Septic tank effluent/sludge	1	1	1	0	2+1	
Sewage plant effluent	3	1	3	1	0	
Sewage plant sludge	3	1	3	0	0	
Soil (topsoil and subsoil)	12	12	14	0	1	39
Forested	2	4	4	0	1	
Fertilizer treated	4	2	4	0	0	
Manure treated	6	6	6	0	0	
Stream water	8	11	18	12	6	55
Forested	1	2	2	1	1	
Fertilizer treated	1+1	1	3+3	3+2	1	
Manure treated	3	3	5	1	0	
Sewage plant effluent	1+1	1+1	1+1	1+1	1+1	
Septic fields	0	2+1	2+1	2+1	1+1	
Runoff	0	0	5	4	0	9
Fertilizer treated	0	0	3	3	0	
Manure treated	0	0	2	1	0	
Precipitation	0	0	7	2	0	9

Table B2. Water-quality field measurements for the Lower Susquehanna River Basin, Pennsylvania

[lat-long-no, latitude, longitude, and site identifier number; µS/cm, microsiemens per centimeter at 25 degrees Celsius; °C, degree Celsius; L/s, liters per second; mg/L, milligrams per liter; --, not measured or data not available; e, estimated value on the basis of observation; ?, measurement attempted, but the value is questionable; <, less than]

Site name	Site number (lat-long-no)	Sample ID	Date	Time	Specific conductance (µS/cm)	pH	Temper-ature (°C)	Discharge rate (L/s)	Suspended sediment (mg/L)
Summer low-flow samples collected June 27 through July 6, 1988									
Bald Eagle Creek (headwaters)	394504-762851-01	BE1WS	880629	1000	--	--	--	Dry	--
Brush Run (at pond)	394911-770625-03	BR2WS	880706	1300	8,100	7.9	32.0	0.057	8.0
Brush Run swine manure	394906-770626-01	BR1MS	880706	1215	12,000	7.8	--	--	--
Codorus Creek at York	395646-764520-01	CCYWS	880630	1115	850	7.2	21.0	1,700	5.0
Codorus Creek at Pleasureville	400107-764136-01	CCPWS	880630	1215	890	6.9	21.5	2,950	10.0
Little Conestoga FS#1 (stream)	400745-755837-01	C1WS	880628	1200	400	7.8	17.5	42	4.0
Little Conestoga FS#2 (spring)	401150-761053-01	C2WG	880628	1400	740	6.8	11.0	1.6	19.0
Monroe Creek #1 (upstream)	402830-762839-01	M1WS	880706	2015	77	6.6	24.0	40	< 1.0
Monroe Creek #2A (downstream)	402810-762944-01	M2AWS	880706	2100	79	7.7	27.0	48	6.0
Stony Creek	402451-764650-01	SCWS	880627	1200	22	5.8	19.0	340	< 1.0
York Wastewater (oxygen)	395917-764327-01	YWOHW	880630	1000	1,000	6.5	24.5	538	5.0
York Wastewater (Contact)	395914-764327-02	YWDHW	880630	1030	1,100	7.3	24.5	765	2.0
Harrisburg Wastewater (oxygen)	401419-765120-01	HWOHW	880630	1500	630	6.3	23.5	1,100	10.0
Winter low-flow samples collected December 14, 1988 through January 5, 1989									
Bald Eagle Creek (at gage)	394454-762750-01	BE2WS	881214	1000	100	7.8	3.0	2.8	< 1.0
Brush Run (at weir)	394906-770626-02	BR1WS	881214	1300	890	7.2	0.5	.031	7.0
Brush Run swine manure	394906-770626-01	BR1MS	881214	1315	--	--	4.5	--	--
Berkshire Hills (upstream)	400824-770139-01	BH1WS	881222	1330	490	8.3	3.0	.062	11.0
Berkshire Hills (downstream)	400836-770140-02	BH2WS	881222	1400	480	8.2	3.0	.062	4.0
Dogwood Run #1 (forested)	400614-770512-01	D1WS	881215	1115	35	8.0	3.0	13	1.0
Dogwood Run #2A (septic)	400653-770230-01	D2AWS	881215	1500	42	7.9	3.0	22	< 1.0
Dogwood Run #3 (mixed/upstream)	400659-770232-01	B3WS	881215	1430	310	8.2	3.0	37	< 1.0
Dogwood Run #4 (sewage/downstream)	400702-770233-01	D4WS	881215	1330	370	7.5	3.5	48	1.0
Little Conestoga FS#1 (stream)	400745-755837-01	C1WS	881221	1200	390	7.8	4.0	82	8.0
Little Conestoga FS#2 (runoff)	401150-761053-01	C2WR	881221	1400	210	7.6	4.0	.031	207.0
Stony Creek (near gate)	402440-764739-01	SCWS	881222	1600	27	8.4	0.5	312	< 1.0
Dillsburg Wastewater	400700-770232-01	DWAHW	881215	1400	570	7.0	10.5	11	3.0

Table B2. Water-quality field measurements for the Lower Susquehanna River Basin, Pennsylvania—Continued

Site name	Site number (lat-long-no)	Sample ID	Date	Time	Specific conductance (µS/cm)	pH	Temper-ature (°C)	Discharge rate (L/s)	Suspended sediment (mg/L)
Spring high-flow samples collected April 27 through May 24, 1989									
Bald Eagle Creek (runoff)	394504-762851-01	BE1WR	890505	1400	150	7.5	15.0	0.14 e	209.0
Bald Eagle Creek (at gage)	394454-762750-02	BE2WS	890505	1300	190	7.4	13.0	144	72.0
Bald Eagle Creek (at gage)	394454-762750-02	BE2WS	890524	1200	120	6.6	--	37	850.0
Brush Run (at weir)	394906-770626-02	BR1WS	890502	1200	--	--	--	156	--
Brush Run (at pond)	394911-770625-03	BR2WS	890505	1030	310	7.4	--	22	54.0
Brush Run swine manure	394906-770626-01	BR1MS	890505	1015	9,200	7.5	--	--	2,260.0
Codorus Creek at York	395646-764520-01	CCYWS	890504	1115	560	7.6	14.0	2,380	17.0
Codorus Creek at Pleasureville	400107-764136-02	CCPWS	890504	1215	540	7.6	14.5	8,300	28.0
Conestoga FS#1 (stream)	400745-755837-01	C1 WS	890427	1000	390	7.7	17.0	116	37.0
Conestoga FS#1 (stream)	400745-755837-01	C1 WS	890502	1100	240	7.1	--	224	150.0
Conestoga FS#1 (runoff)	400742-755840-01	C1 WR	890502	1100	160	7.3	--	8.5e	1770.0
Conestoga FS#2 (spring)	401150-761053-01	C2 WG	890428	0900	720	7.5	--	.31	52.0
Conestoga FS#2 (runoff 1)	401150-761053-01	C2AWR	890502	0930	200	7.2	--	.14 e	29.0
Conestoga FS#2 (runoff 2)	401156-761100-01	C2BWR	890502	0930	2,200	7.4	--	.14 e	6,710.0
Berkshire Hills #1 (upstream)	400824-770139-01	BH1WS	890503	0900	410	7.9	12.0	1.3 e	10.0
Berkshire Hills #2 (downstream)	400836-770140-01	BH2WS	890503	0915	420	8.2	12.0	1.4 e	6.0
Berkshire Hills #1 (upstream)	400824-770139-01	BH1WS	890518	0900	340	7.8	13.0	1.3 e	25.0
Berkshire Hills #2 (downstream)	400836-770140-01	BH2WS	890518	0915	330	7.7	13.0	1.4 e	14.0
Dogwood Run #1 (forested)	400614-770512-01	D1 WS	890503	1000	30	7.5	14.0	74	7.0
Dogwood Run #2A (septic)	400653-770230-01	D2AWS	890503	1100	93	6.9	14.5	116	6.0
Dogwood Run #3 (mixed/upstream)	400659-770232-01	D3 WS	890503	1230	190	7.2	14.0	195	9.0
Dogwood Run #4 (sewage/downstream)	400702-770233-01	D4 WS	890503	1330	230	7.2	14.0	212	8.0
Dogwood Run #1 (forested)	400614-770512-01	D1 WS	890518	1000	33	5.9	15.5	538	8.0
Dogwood Run #2A (septic)	400653-770230-01	D2AWS	890518	1100	47	6.8	16.0	793 e	30.0
Dogwood Run #2B (golf course)	400653-770229-01	D2BWS	890518	1200	380	7.8	14.0	181	22.0
Dogwood Run #3 (mixed/upstream)	400659-770232-01	D3 WS	890518	1330	120	7.0	14.0	991	17.0
Dogwood Run #4 (sewage/downstream)	400702-770233-01	D4 WS	890518	1430	190	7.2	14.0	1,080	16.0
Monroe Creek #1 (upstream)	402831-762839-01	M1 WS	890510	1330	56	6.8	10.5	1,220	10.0
Monroe Creek #2A (midstream)	402810-762944-01	M2AWS	890510	1230	67	6.7	10.5	1,560	18.0
Monroe Creek #3 (downstream)	402810-763003-01	M3 WS	890510	1130	68	6.9	10.5	1,730	16.0
Monroe Creek #2A (runoff)	402810-762944-02	M2AWR	890510	1100	60	7.7	10.5	.14 e	5.0
Monroe Creek #2C (spring)	402811-763003-01	M2CWG	890510	1030	200	6.7	11.0	.085 e	3.0
Monroe Creek #1 (upstream)	402831-762839-01	M1 WS	890517	1330	53	7.1	14.5	1,810	22.0
Monroe Creek #2A (midstream)	402810-762944-01	M2AWS	890517	1230	57	7.0	14.0	1,980	18.0
Monroe Creek #2B (tributary)	402811-762958-01	M2BWS	890517	1100	92	6.8	12.5	340	27.0
Monroe Creek #3 (downstream)	402810-763003-01	M3 WS	890517	1000	61	6.9	14.5	2,350 e	21.0
Monroe Creek #2C (spring)	402811-763003-01	M2CWG	890517	0930	--	--	--	--	--
Stony Creek	402440-764730-01	SC WS	890503	0830	26	7.8	11.0	2,010	3.0
Dillsburg Wastewater	400700-770232-01	DWAHW	890503	1400	470	7.3	14.0	18	3.0
Dillsburg Wastewater	400700-770232-01	DWAHW	890518	1400	--	--	--	82	--
York Wastewater	395917-764327-02	YWRHW	890504	0930	1,000	7.4	--	793 e	5.0
Harrisburg Wastewater	401419-765120-01	HWOHW	890504	1500	600	7.4	--	793 e	7.0
Harrisburg septic-tank	401419-765120-02	HW HS	890505	1100	--	--	--	--	--

Table B2. Water-quality field measurements for the Lower Susquehanna River Basin, Pennsylvania—Continued

Site name	Site number (lat-long-no)	Sample ID	Date	Time	Specific conductance (µS/cm)	pH	Temper-ature (°C)	Discharge rate (L/s)	Suspended sediment (mg/L)
Summer high-flow samples collected June 9 through July 6, 1989									
Bald Eagle Creek (runoff)	394504-762851-01	BE1WR	890609	1150	--	--	--	0.14 e	--
Conestoga FS#1 (stream)	400745-755837-01	C1 WS	890705	1100	--	--	--	227 e	
Conestoga FS#1 (runoff)	400742-755840-02	C1 WR	890705	1100	--	--	--	28 e	200.0
Dogwood Run #1 (forested)	400614-770512-01	D1 WS	890706	1435	--	--	19.0	312	8.0
Dogwood Run #2A (septic)	400653-770230-01	D2AWS	890706	1240	--	--	18.1	425	13.0
Dogwood Run #2B (golf course)	400653-770229-01	D2BWS	890706	1230	--	--	19.0	170	16.0
Dogwood Run #3 (mixed/upstream)	400659-770232-01	D3 WS	890706	1330	--	--	18.1	595	14.0
Dogwood Run #4 (sewage/downstream)	400702-770233-01	D4 WS	890706	1430	--	--	18.0	680	14.0
Monroe Creek #1 (upstream)	402831-762839-01	M1 WS	890706	1330	56	7.0	--	765	9.0
Monroe Creek #2A (midstream)	402810-762944-01	M2AWS	890706	1230	66	7.1	--	878	9.0
Monroe Creek #2B (tributary)	402811-762958-01	M2BWS	890706	1100	150	6.8	19.5	116	12.0
Monroe Creek #3 (downstream)	402810-763003-01	M3 WS	890706	1015	74	7.1	18.0	991	11.0
Monroe Creek #2A (runoff)	402810-762944-02	M2AWR	890706	1145	56	7.2	25.5	.14 e	4.0
Monroe Creek #2C (spring)	402811-763003-01	M2CWG	890706	1030	340	7.3	22.0	.085 e	18.0
Dillsburg Wastewater	400700-770232-01	DWAWH	890706	1010	--	--	17.5	65	6.0
Fall and winter low-flow samples collected October 12, 1989, and January 26, 1990									
Dogwood Run #1 (forested)	400614-770512-01	D1 WS	891012	1300	28	8.3	13.0	17	6.0
Dogwood Run #2A (septic)	400653-770230-01	D2AWS	891012	1130	72	7.8	10.5	28	2.0
Dogwood Run #2B (golf course)	400653-770229-01	D2BWS	891012	1030	510	7.8	11.5	15 e	1.0
Dogwood Run #3 (mixed/upstream)	400659-770232-01	D3 WS	891012	0930	260	7.9	11.0	57	1.0
Dogwood Run #4 (sewage/downstream)	400702-770233-01	D4 WS	891012	0830	330	7.7	12.0	74	2.0
Dillsburg Wastewater	400700-770232-01	DWAHW	890706	0900	520	7.4	17.0	17	7.0
Berks well water (septic)	403051-761115-01	BK1WG	900126	0900	160	7.6	10.0	--	<1.0
Berks septic effluent (septic)	403050-761114-02	BK2HS	900126	0900	590	7.1	11.0	--	3.0
Bulk precipitation-quality data for spring and summer high-flow samples collected May 2 through 18 and July 5 through 7, 1989									
								Rainfall, in centimeters per 24 hours	
Bald Eagle Creek (precipitation)	394504-762851-01	BE1WP	890502		27	4.5	--	4.34	--
Brush Run (precipitation)	394906-770626-02	BR1WP	890502		23	6.5	--	4.47	--
Monroe Creek (precipitation)	402810-762944-01	M2AWP	890502		13	4.8	--	3.15	--
Monroe Creek (precipitation)	402810-762944-01	M2AWP	890510		34	4.4	--	1.50	--
Monroe Creek (precipitation)	402810-762944-01	M2AWP	890705		19	4.6	--	1.20 e	--
Conestoga FS#1+2 (precipitation)	(combined)	C12WP	890502		20	5.0	--	?	--
Conestoga FS#1 (precipitation)	400742-755840-01	C1 WP	890505		20	4.6	--	4.14	--
Conestoga FS#1 (precipitation)	400742-755840-01	C1 WP	890705		17	5.0	--	1.5?	--
Conestoga FS#2 (precipitation)	400742-755840-01	C2 WP	890505		12	4.8	--	4.22	--

Table B3. Concentrations of carbon (C), nitrogen (N), and sulfur (S) in N-source and fluvial samples from selected small watersheds in the Lower Susquehanna River Basin, Pennsylvania

[Concentrations are in parts per million (aqueous solutes are in milligrams per liter; solid compounds are in milligrams per kilogram); --, no data; <, less than; Replicate analysis results indicated by slash between values (for example, 1,120/1,180)]

Sample ID	Date sampled	Sample location and description	Carbon as C		Nitrogen as N				Sulfur as S		
			C-inorg	C-org	NO_3	NO_2	NH_3	N-org	Sulfate	Sulfide	S-org
		Samples collected during June 27 through July 6, 1988									
BE1FL	880629	Bald Eagle synthetic fertilizer (liquid 30 percent N)	27	94,000	110,000	0.1	153,000	90,000	265	1.0	--
BE1FS	880629	Bald Eagle synthetic fertilizer (solid 15-15-15)	200	21,600	280	--	127,670	17,730	99,700	<100	300
BE1S1	880629	Bald Eagle topsoil	600	15,700	390	--	780	1,820	300	<100	<100
BE1S2	880629	Bald Eagle subsoil	700	13,900	180	--	290	1,180	100	<100	<100
BR1MC	880706	Brush Run chicken manure	1,700	334,700	1,150	--	4,430	39,230	3,500	1,600	100
BR1MS1	880706	Brush Run swine manure, filtered (liquid)	--	--	6.0	<.1	1,500	125	10.0	.8	--
BR1MS2	880706	Brush Run swine manure, whole (liquid)	230	370	5.0	<.1	1,400	800	8.0	.8	--
BR1S1	880706	Brush Run topsoil	600	31,800	280	--	480	900	200	<100	<100
BR1S2	880706	Brush Run subsoil	400	18,600	130	--	290	2,460	100	<100	<100
BR2WS1	880706	Brush Run stream water (pond), filtered	--	--	.2	<.1	4.8	2.4	44.0	.1	--
BR2WS2	880706	Brush Run stream water (pond), whole	40	19	.5	<.1	5.4	3.5	42.3	.1	--
CCPWS1	880630	Codorus Creek Pleasureville stream water, filtered	--	--	2.0	.2	1.4	2.4	34.7	<.1	--
CCPWS2	880630	Codorus Creek Pleasureville stream water, whole	1.8	15	2.0	.1	1.1	2.9	32.3	<.1	--
CCYWS1	880630	Codorus Creek York stream water, filtered	--	--	1.0	<.1	.3	2.4	34.3	.1	--
CCYWS2	880630	Codorus Creek York stream water, whole	12	16	1.1	<.1	.3	2.2	33.3	.1	--
C1MD	880628	Conestoga FS#1 dairy manure	16,300	330,500	1,060	--	6,390	17,270	3,200	1,500	<100
C1S1	880628	Conestoga FS#1 topsoil	1,200	17,400	220	--	290	1,770	<100	<100	<100
C1S2	880628	Conestoga FS#1 subsoil	1,000	10,000	200	--	190	1,050	<100	<100	<100
C1WS1	880628	Conestoga FS#1 stream water, filtered	--	--	4.9	<.1	.3	.7	9.3	.1	--
C1WS2	880628	Conestoga FS#1 stream water, whole	17	1.7	5.1	<.1	.2	1.0	9.3	.1	--

Table B3. Concentrations of carbon (C), nitrogen (N), and sulfur (S) in N-source and fluvial samples from selected small watersheds in the Lower Susquehanna River Basin, Pennsylvania—Continued

[Concentrations are in parts per million (aqueous solutes are in milligrams per liter; solid compounds are in milligrams per kilogram); --, no data; <, less than; Replicate analysis results indicated by slash between values (for example, 1,120/1,180)]

Sample ID	Date sampled	Sample location and description	Carbon as C		Nitrogen as N				Sulfur as S		
			C-inorg	C-org	NO_3	NO_2	NH_3	N-org	Sulfate	Sulfide	S-org
Samples collected during June 27 through July 6, 1988—Continued											
C2 MB	880628	Conestoga FS#2 steer manure	300	370,600	1,250	--	7,760	24,100	100	3,300	<100
C2 MS	880628	Conestoga FS#2 swine manure	2,400	333,900	1,230	--	17,950	15,170	400	6,700	100
C2 MC	880628	Conestoga FS#2 chicken manure	900	389,800	3,160	--	6,160	37,150	1,900	3,400	<100
C2 S1	880628	Conestoga FS#2 topsoil	1,500	29,100	290	--	350	2,260	200	<100	<100
C2 S2	880628	Conestoga FS#2 subsoil	700	16,500	290	--	280	1,410	<100	<100	<100
C2 WS1	880628	Conestoga FS#2 spring water, filtered	--	--	16	<.1	.2	.5	17.3	<.1	--
C2 WS2	880628	Conestoga FS#2 spring water, whole	32	.8	15	<.1	.1	.4	16.7	<.1	--
M2AFS	880706	Monroe Creek synth. fertilizer (solid 34-6-7)	400	67,900	3,730	--	46,080	43,500	6,000	2,200	1,900
M2AFS	D880706	Monroe Creek synth. fertilizer (solid 34-6-7)	30,600	46,100	3,800	--	32,300	40,300	5,670	4,430	<100
M2AS1	880706	Monroe Creek topsoil	900	77,700	220	--	880	5,140	300	<100	<100
M2AS2	880706	Monroe Creek subsoil	1,000	21,600	60	--	480	1,760	200	<100	<100
M1 WS1	880706	Monroe Creek #1 upstream water, filtered	--	--	.8	<.1	.6	.9	.7	.1	--
M1 WS2	880706	Monroe Creek #1 upstream water, whole	4.5	2.3	.5	<.1	.9	.9	.7	.1	--
M2AWS1	880706	Monroe Creek #2A midstream water, filtered	--	--	.5	<.1	.1	1.4	.3	.1	--
M2AWS2	880706	Monroe Creek #2A midstream water, whole	3.9	2.5	.6	<.1	.2	1.5	.7	.1	--
SC S1	880627	Stony Creek topsoil	1,000	163,200	290	--	690	5,260	200	200	100
SC S2	880627	Stony Creek subsoil	900	25,100	50	--	10	1,340	100	<100	<100
SC WS1	880627	Stony Creek stream water, filtered	--	--	.5	<.1	.3	1.1	.7	.1	--
SC WS2	880627	Stony Creek stream water, whole	.9	1.5	.5	<.1	.4	1.2	.3	.1	--
HWOHW1	880630	Harrisburg Wastewater Plant O_2 effluent, filtered	--	--	.8	<.1	19.0	<.1	22.7	.1	--
HWOHW2	880630	Harrisburg Wastewater Plant O_2 effluent, whole	23	11	.6	<.1	15.2	8.4	21.0	<.1	--
HWOHW6	880630	Harrisburg Wastewater Plant primary sludge	3,800	259,600	110	--	6,890	24,290	300	4,100	1,100
HW HS6	880630	Harrisburg Wastewater Plant septic tank sludge	2,300	214,200	10	--	9,000	24,270	2,300	3,300	1,100
YW HW6	880630	York Wastewater Plant primary sludge	1,800	303,800	10	--	14,170	31,910	900	6,900	600

Table B3. Concentrations of carbon (C), nitrogen (N), and sulfur (S) in N-source and fluvial samples from selected small watersheds in the Lower Susquehanna River Basin, Pennsylvania—Continued

[Concentrations are in parts per million (aqueous solutes are in milligrams per liter; solid compounds are in milligrams per kilogram); --, no data; <, less than; Replicate analysis results indicated by slash between values (for example, 1,120/1,180)]

Sample ID	Date sampled	Sample location and description	Carbon as C		Nitrogen as N				Sulfur as S		
			C-inorg	C-org	NO_3	NO_2	NH_3	N-org	Sulfate	Sulfide	S-org
		Samples collected during December 14, 1988, through January 5, 1989									
BE1S1	881214	Bald Eagle topsoil	4,200	21,200	111	--	171	2,105	400	<100	400
BE1S2	881214	Bald Eagle subsoil	200	14,500	53	--	152	1,392	600	<100	100
BE2WS1	881214	Bald Eagle downstream water, filtered	--	--	.7	<.1	<.1	.5	<1.8	<.1	--
BE2WS2	881214	Bald Eagle downstream water, whole	<5	.9	.7	<.1	<.1	1.0	<1.8	<.1	--
BH1WS1	881222	Berkshire Hills #1 septic upstream, filtered	--	--	1.3	<.1	<0.1	.5	9.7	<.1	--
BH1WS2	881222	Berkshire Hills #1 septic upstream, whole	45	1.6	1.2	<.1	<0.1	.5	9.3	<.1	--
BH2WS1	881222	Berkshire Hills #2 septic downstream, filtered	--	--	.8	<.1	<.1	.6	6.0	<.1	--
BH2WS2	881222	Berkshire Hills #2 septic downstream, whole	46	1.4	1.2	<.1	<.1	1.6	9.0	<.1	--
BR1MS4	881214	Brush Run swine feces	<100	400,000	987	--	5,860	23,410	700	400	1,400
BR1MS1	881214	Brush Run swine manure, filtered	--	--	.6/.2	.1/<.1	1,120/1,180	160/30	<1.8/<1.8	.01	--
BR1MS2	881214	Brush Run swine manure, whole	37	870	.9/.2	.1/<.1	1,140/1,190	164/100	<1.8/<1.8	.	--
BR1S1	881214	Brush Run topsoil	<100	37,700	203	--	304	3,753	600	<100	500
BR1S2	881214	Brush Run subsoil	<100	22,500	115	--	202	2,202	200	<100	200
BR1WS1	881214	Brush Run stream water (weir), filtered	--	--	4.7	<.1	3.7	.2	29.0	<.1	--
BR1WS2	881214	Brush Run stream water (weir), whole	66	11	4.8	<.1	4.4	.4	29.7	<.1	--
D1 S1	881215	Dogwood Run #1 topsoil	1,300	497,100	426	--	1,384	19,678	500	<100	300
D1 S2	881215	Dogwood Run #1 subsoil	<100	20,400	77	--	77	1,189	400	<100	400
D1 WS1	881215	Dogwood Run #1 upstream water, filtered	--	--	<.1	<.1	<.1	.9	<1.8	<.1	--
D1 WS2	881215	Dogwood Run #1 upstream water, whole	<5	3.6	<.1	<.1	<.1	.7	<1.8	<.1	--
D2AWS1	881215	Dogwood Run #2A septic downstream, filtered	--	--	.1	<.1	<.1	.5	<1.8	<.1	--
D2AWS2	881215	Dogwood Run #2A septic downstream, whole	<5	2.4	<.1	<.1	<.1	.4	<1.8	<.1	--
D3 WS1	881215	Dogwood Run #3 mixed downstream, filtered	--	--	.4	<.1	.1	.4	1.8	<.1	--
D3 WS2	881215	Dogwood Run #3 mixed downstream, whole	25	1.9	1.5	<.1	<.1	.5	7.1	<.1	--
D4 WS1	881215	Dogwood Run #4 sewage downstream, filtered	--	--	4.4	<.1	.1	.4	13.1	<.1	--
D4 WS2	881215	Dogwood Run #4 sewage downstream, whole	24	2.4	4.3	<.1	<.1	.7	13.1	<.1	--

Table B3. Concentrations of carbon (C), nitrogen (N), and sulfur (S) in N-source and fluvial samples from selected small watersheds in the Lower Susquehanna River Basin, Pennsylvania—Continued

[Concentrations are in parts per million (aqueous solutes are in milligrams per liter; solid compounds are in milligrams per kilogram); --, no data; <, less than; Replicate analysis results indicated by slash between values (for example, 1,120/1,180)]

Sample ID	Date sampled	Sample location and description	Carbon as C		Nitrogen as N				Sulfur as S		
			C-inorg	C-org	NO_3	NO_2	NH_3	N-org	Sulfate	Sulfide	S-org
Samples collected during December 14, 1988, through January 5, 1989—Continued											
CLMMD	881221	Conestoga LM dairy manure	13,300	363,000	681	--	11,284	19,248	1,500	1,500	<100
C1 MD	881221	Conestoga FS#1 dairy manure	31,400	303,200	634	--	3,609	15,264	100	2,100	200
C1 S1	881221	Conestoga FS#1 topsoil	1,300	26,600	113	--	272	2,157	200	<100	<100
C1 S2	881221	Conestoga FS#1 subsoil	700	19,000	111	--	177	1,567	1,100	<100	<100
C1 WS1	881221	Conestoga FS#1 stream water, filtered	--	--	5.2	0.1	<.1	.4	4.3	<.1	--
C1 WS2	881221	Conestoga FS#1 stream water, whole	33	1.00	5.6	.1	.1	.6	5.3	<.1	--
C2 MB	881221	Conestoga FS#2 steer manure	<100	421,800	775	--	4,078	29,552	2,500	300	200
C2 MC	881221	Conestoga FS#2 chicken manure	<100	399,900	1,557	--	12,909	65,954	100	4,300	100
C2 MS1	881221	Conestoga FS#2 swine manure, filtered	--	--	1.2	<.1	2,780	1,540	139.9	<.1	--
C2 MS2	881221	Conestoga FS#2 swine manure, whole	190	8,400	1.1	<.1	2,920	1,320	143.4	<.1	--
C2 S1	881221	Conestoga FS#2 topsoil	300	23,300	123	--	383	1,790	1,100	<100	<100
C2 S2	881221	Conestoga FS#2 subsoil	<100	17,900	89	--	202	1,442	100	<100	<100
C2 WR1	881221	Conestoga FS#2 runoff water, filtered	--	--	.6	<.1	3.5	4.1	6.7	<.1	--
C2 WR2	881221	Conestoga FS#2 runoff water, whole	23	26	.6	<.1	3.7	4.8	7.3	<.1	--
SC S1	881222	Stony Creek topsoil	6,500	231,900	271	--	985	8,405	200	300	200
SC S2	881222	Stony Creek subsoil	<100	37,500	14	--	58	1,412	100	<100	<100
SC WS1	881222	Stony Creek stream water, filtered	--	--	<.1	<.1	.1	.8	<1.8	<.1	--
SC WS2	881222	Stony Creek stream water, whole	<5	1.10	<.1	<.1	.1	.5	<1.8	<.1	--
DWAHW1	881215	Dillsburg Wastewater Plant effluent, filtered	--	--	8.2	<.1	<.1	.7	17.7	<.1	--
DWAHW2	881215	Dillsburg Wastewater Plant effluent, whole	17.4	3.90	8.2	<.1	<.1	.7	18.0	<.1	--
DWAHW4	881215	Dillsburg Wastewater Plant sludge, filtered	--	--	12.3	<.1	15.3	<.1	10.6	<.1	--
DWAHW5	881215	Dillsburg Wastewater Plant sludge, whole	13.0	440	24.8	<.1	11.5	917	42.5	<.1	--
HW HS6	890105	Harrisburg Wastewater Plant septic tank sludge	--	--	--	--	--	--	--	--	--
HW HS1	890105	Harrisburg Wastewater Plant septic tank, filtered	--	--	<.1	<.1	102	<.1	19.1	<.1	--
HW HS2	890105	Harrisburg Wastewater Plant septic tank, whole	1,100	5,900	<.1	.1	112	560	18.1	1.7	--

Table B3. Concentrations of carbon (C), nitrogen (N), and sulfur (S) in N-source and fluvial samples from selected small watersheds in the Lower Susquehanna River Basin, Pennsylvania—Continued

[Concentrations are in parts per million (aqueous solutes are in milligrams per liter; solid compounds are in milligrams per kilogram); --, no data; <, less than; Replicate analysis results indicated by slash between values (for example, 1,120/1,180)]

Sample ID	Date sampled	Sample location and description	Carbon as C		Nitrogen as N				Sulfur as S		
			C-inorg	C-org	NO_3	NO_2	NH_3	N-org	Sulfate	Sulfide	S-org
		Samples collected during April 27 through May 24, 1989									
BE1FS	890505	Bald Eagle synth. fertilizer (solid 15-15-15)	1,500	27,200	500	--	124,600	700	67,500	21,200	200
BE1S1	890505	Bald Eagle topsoil	1,600	15,800	70	<0.1	200	1,400	200	<100	<100
BE1S2	890505	Bald Eagle subsoil	<100	19,200	70	<.1	250	1,310	200	<100	<100
BE1WR1	890505	Bald Eagle runoff water, filtered	--	--	.8	<.1	.1	.6	1.3	<.1	--
BE1WR2	890505	Bald Eagle runoff water, whole	18	4.5	4.2	<.1	2.8	1.2	1.0	<.1	--
BE2WS1	890505	Bald Eagle stream water, filtered	--	--	1.7	<.1	.3	.9	1.3	<.1	--
BE2WS2	890505	Bald Eagle stream water, whole	11	13	1.2	<.1	.3	1.4	1.3	<.1	--
BE2WS1	890524	Bald Eagle stream water, filtered	--	--	2.4	<.1	.1	1.0	2.3	<.1	--
BE2WS2	890524	Bald Eagle stream water, whole	9.0	5.7	2.4	<.1	.2	1.3	2.0	<.1	--
BH1WS1	890518	Berkshire Hills #1 septic upstream, filtered	--	--	.5	<.1	.1	.6	7.7	<.1	--
BH1WS2	890518	Berkshire Hills #1 septic upstream, whole	34	3.4	.8	<.1	<.1	.6	6.7	<.1	--
BH2WS1	890518	Berkshire Hills #2 septic downstream, filtered	--	--	.9	<.1	<.1	.7	8.3	<.1	--
BH2WS2	890518	Berkshire Hills #2 septic downstream, whole	35	3.3	.8	<.1	<.1	.7	6.7	<.1	--
BR1MS1	890505	Brush Run swine manure, filtered	--	--	1.1	<.1	776	16.	1.7	.1	--
BR1MS2	890505	Brush Run swine manure, whole	1,200	913	3.3	<.1	800	16.	1.7	.3	--
BR1S1	890505	Brush Run topsoil	1,000	22,400	90	--	450	2,540	200	<100	<100
BR1S2	890505	Brush Run subsoil	60	14,600	90	--	800	1,400	200	<100	<100
BR2WS1	890505	Brush Run stream water, filtered	--	--	3.9	<.1	.1	3.8	9.7	<.1	--
BR2WS2	890505	Brush Run stream water, whole	19	24	2.2	<.1	1.9	2.6	5.7	<.1	--

Table B3. Concentrations of carbon (C), nitrogen (N), and sulfur (S) in N-source and fluvial samples from selected small watersheds in the Lower Susquehanna River Basin, Pennsylvania—Continued

[Concentrations are in parts per million (aqueous solutes are in milligrams per liter; solid compounds are in milligrams per kilogram); --, no data; <, less than; Replicate analysis results indicated by slash between values (for example, 1,120/1,180)]

Sample ID	Date sampled	Sample location and description	Carbon as C		Nitrogen as N				Sulfur as S		
			C-inorg	C-org	NO_3	NO_2	NH_3	N-org	Sulfate	Sulfide	S-org
Samples collected during April 27 through May 24, 1989—Continued											
D1 S1	890503	Dogwood Run #1 topsoil	400	155,100	150	<0.1	680	5,830	300	300	100
D1 S2	890503	Dogwood Run #1 subsoil	400	33,700	60	<.1	190	1,120	200	<100	<100
D1 WS1	890518	Dogwood Run #1 upstream water, filtered	--	--	<.1	<.1	<.1	.5	4.0	<.1	--
D1 WS2	890518	Dogwood Run #1 upstream water, whole	1.0	4.2	<.1	<.1	<.1	.5	4.3	<.1	--
D2AWS1	890518	Dogwood Run #2A septic downstream, filtered	--	--	.1	<.1	<.1	.4	4.7	<.1	--
D2AWS2	890518	Dogwood Run #2A septic downstream, whole	2.0	4.2	.2	<.1	<.1	.6	4.7	<.1	--
D2BWS1	890518	Dogwood Run #2B golf course trib, filtered	--	--	1.8	<.1	<.1	.7	18.3	<.1	--
D2BWS2	890518	Dogwood Run #2B golf course trib, whole	34	3.4	1.8	<.1	<.1	.8	18.3	<.1	--
D3 WS1	890518	Dogwood Run #3 mixed downstream, filtered	--	--	.4	<.1	<.1	.7	6.3	<.1	--
D3 WS2	890518	Dogwood Run #3 mixed downstream, whole	8.0	4.1	.5	<.1	<.1	.6	7.0	<.1	--
D4 WS1	890518	Dogwood Run #4 sewage downstream, filtered	--	--	1.3	<.1	.1	.7	8.7	<.1	--
D4 WS2	890518	Dogwood Run #4 sewage downstream, whole	14	3.9	1.2	<.1	<.1	1.3	7.0	<.1	--
SC S1	890503	Stony Creek topsoil	400	189,900	170	--	670	5,930	200	300	100
SC S2	890503	Stony Creek subsoil	300	24,900	120	--	400	720	200	<100	<100
SC WS1	890503	Stony Creek stream water, filtered	--	--	<.1	<.1	.2	.5	2.0	<.1	--
SC WS2	890503	Stony Creek stream water, whole	1.2	3.7	<.1	<.1	.3	.5	1.7	<.1	--
CCPWS1	890504	Codorus Creek Pleasureville stream water, filtered	--	--	2.6	<.1	.1	.9	11.7	<.1	--
CCPWS2	890504	Codorus Creek Pleasureville stream water, whole	7.6	22	2.5	<.1	.1	.9	11.3	<.1	--
CCYWS1	890504	Codorus Creek York stream water, filtered	--	--	1.9	<.1	.1	.9	9.7	<.1	--
CCYWS2	890504	Codorus Creek York stream water, whole	18	11	1.7	<.1	.1	.9	8.0	<.1	--
C1 MD	890427	Conestoga FS#1 dairy manure	2,900	297,700	600	--	18,990	5,960	100	4,100	100
C1 S1	890427	Conestoga FS#1 topsoil	600	13,000	60	--	170	1,030	200	<100	<100
C1 S2	890427	Conestoga FS#1 subsoil	700	12,000	50	--	170	960	100	<100	<100
C1 WR1	890502	Conestoga FS#1 runoff water, filtered	--	--	.5	<.1	.4	1.2	3.7	<.1	--
C1 WR2	890502	Conestoga FS#1 runoff water, whole	16	10	.6	<.1	.6	1.8	3.7	<.1	--
C1 WS1	890502	Conestoga FS#1 stream water (high), filtered	--	--	1.3	<.1	.1	.7	2.7	<.1	--
C1 WS2	890502	Conestoga FS#1 stream water (high), whole	20	5.3	1.1	<.1	.1	1.5	1.7	<.1	--
C1 WS1	890427	Conestoga FS#1 stream water (low), filtered	--	--	4.2	<.1	<.1	.4	6.3	<.1	--
C1 WS2	890427	Conestoga FS#1 stream water (low), whole	1.8	30	1.3	<.1	.1	.5	1.3	--	--

Table B3. Concentrations of carbon (C), nitrogen (N), and sulfur (S) in N-source and fluvial samples from selected small watersheds in the Lower Susquehanna River Basin, Pennsylvania—Continued

[Concentrations are in parts per million (aqueous solutes are in milligrams per liter; solid compounds are in milligrams per kilogram); --, no data; <, less than; Replicate analysis results indicated by slash between values (for example, 1,120/1,180)]

Sample ID	Date sampled	Sample location and description	Carbon as C		Nitrogen as N				Sulfur as S		
			C-inorg	C-org	NO_3	NO_2	NH_3	N-org	Sulfate	Sulfide	S-org
Samples collected during April 27 through May 24, 1989—Continued											
C2 MB	890428	Conestoga FS#2 steer manure	700	345,400	600	--	19,440	8,960	400	2,900	100
C2 MC	890428	Conestoga FS#2 chicken manure	1,000	325,700	770	--	26,850	52,410	900	3,100	200
C2 MS1	890428	Conestoga FS#2 swine manure, filtered	--	--	.7	<0.1	3,500	400	6.7	.1	--
C2 MS2	890428	Conestoga FS#2 swine manure, whole	37	20,000	.5	<.1	4,200	300	45.7	.2	--
C2 S1	890428	Conestoga FS#2 topsoil	600	12,100	60	--	760	550	<100	<100	<100
C2 S2	890428	Conestoga FS#2 subsoil	300	9,700	80	--	500	760	300	<100	<100
C2 WG1	890428	Conestoga FS#2 spring water, filtered	--	--	3.0	<.1	<.1	.3	6.7	<.1	--
C2 WG2	890428	Conestoga FS#2 spring water, whole	62	.9	3.3	<.1	<.1	.3	3.0	<.1	--
C2AWR1	890502	Conestoga FS#2 runoff #1 water, filtered	--	--	<.1	<.1	.1	2.0	1.6	<.1	--
C2AWR2	890502	Conestoga FS#2 runoff #1 water, whole	27	41.	<.1	<.1	.1	3.3	1.3	<.1	--
C2BWR1	890502	Conestoga FS#2 runoff #2 water, filtered	--	--	.6	<.1	51.8	36.2	10	<.1	--
C2BWR2	890502	Conestoga FS#2 runoff #2 water, whole	150	380	2.0	<.1	73.8	46.2	20	<.1	--
M2AFS1	890510	Monroe Creek synth. fertilizer (solid 34-3-8)	2,500	24,170	24,200	--	15,300	300,700	9,950	15,950	37,400
M2AFS2	890510	Monroe Creek synth. fertilizer (solid 18-4-10)	17,500	184,900	8,100	--	26,800	145,200	48,500	10,700	<100
M2AS1	890502	Monroe Creek topsoil	300	44,000	100	--	990	2,490	300	200	<100
M2AS2	890502	Monroe Creek subsoil	800	21,000	110	--	500	1,930	100	<100	<100
M1 WS1	890510	Monroe Creek #1 upstream water, filtered	--	--	.5	<.1	<.1	.5	2.3	<.1	--
M1 WS2	890510	Monroe Creek #1 upstream water, whole	4.0	2.7	.4	<.1	<.1	.5	2.0	<.1	--
M2AWS1	890510	Monroe Creek #2A midstream water, filtered	--	--	.7	<.1	<.1	.5	3.0	.2	--
M2AWS2	890510	Monroe Creek #2A midstream water, whole	5.0	3.3	.7	<.1	<.1	1.1	3.3	<.1	--
M3 WS1	890510	Monroe Creek #3 downstream water, filtered	--	--	.7	<.1	<.1	1.1	3.3	<.1	--
M3 WS2	890510	Monroe Creek #3 downstream water, whole	5.0	4.0	.5	<.1	<.1	1.1	2.0	<.1	--
M2AWR1	890510	Monroe Creek #2A runoff, filtered	--	--	<.1	<.1	.1	.7	1.7	<.1	--
M2AWR2	890510	Monroe Creek #2A runoff, whole	8.0	20.0	<.1	<.1	.1	1.2	2.3	<.1	--
M2CWG1	890510	Monroe Creek #2C spring, filtered	--	--	.6	<.1	<.1	.5	4.7	<.1	--
M2CWG2	890510	Monroe Creek #2C spring, whole	15.0	3.5	.4	<.1	<.1	.6	2.7	<.1	--

Table B3. Concentrations of carbon (C), nitrogen (N), and sulfur (S) in N-source and fluvial samples from selected small watersheds in the Lower Susquehanna River Basin, Pennsylvania—Continued

[Concentrations are in parts per million (aqueous solutes are in milligrams per liter; solid compounds are in milligrams per kilogram); --, no data; <, less than; Replicate analysis results indicated by slash between values (for example, 1,120/1,180)]

Sample ID	Date sampled	Sample location and description	Carbon as C		Nitrogen as N				Sulfur as S		
			C-inorg	C-org	NO_3	NO_2	NH_3	N-org	Sulfate	Sulfide	S-org
		Samples collected during April 27 through May 24, 1989—Continued									
DWAHW1	890518	Dillsburg Wastewater Plant effluent, filtered	--	--	2.1	<0.1	<0.1	0.5	9.7	<0.1	--
DWAHW2	890518	Dillsburg Wastewater Plant effluent, whole	<1.0	3.5	2.1	<.1	<.1	.8	9.3	<.1	--
HW HS1	890505	Harrisburg Wastewater Plant septic tank, filtered	--	--	<.1	<.1	109	1.2	1.0	<.1	--
HW HS2	890505	Harrisburg Wastewater Plant septic tank, whole	14	390	.2	<.1	164	24.4	4.7	<.1	--
HWOHW1	890504	Harrisburg Wastewater Plant effluent, filtered	--	--	.1	<.1	8.1	1.2	9.0	<.1	--
HWOHW2	890504	Harrisburg Wastewater Plant effluent, whole	34	8.9	.1	<.1	13.3	2.6	9.3	<.1	--
YWRHW1	890504	York Wastewater Plant effluent, filtered	--	--	3.9	<.1	.3	1.1	37.0	<.1	--
YWRHW2	890504	York Wastewater Plant effluent, whole	37	12	7.0	<.1	.3	1.4	39.3	<.1	--
BE2WP1	890502	Bald Eagle precipitation, filtered	--	--	.1	<.1	.1	.6	.3	<.1	--
BR1WP1	890502	Brush Run precipitation, filtered	--	--	.3	<.1	1.4	.3	1.0	<.1	--
C1 WP1	890505	Conestoga FS#1 precipitation, filtered	--	--	<.1	<.1	.2	1.2	<.3	<.1	--
C2 WP1	890505	Conestoga FS#2 precipitation, filtered	--	--	<.1	<.1	.1	1.2	<.3	<.1	--
M2AWP1	890502	Monroe Creek precipitation, filtered	--	--	.1	<.1	.1	.6	<.3	<.1	--
M2AWP1	890510	Monroe Creek precipitation, filtered	--	--	.3	<.1	.1	.6	.3	<.1	--
		Samples collected during June 9 through July 6, 1989									
D1 WS1	890706	Dogwood Run #1 upstream water, filtered	--	--	<.1	<.1	<.1	.7	2.0	<.1	--
D1 WS2	890706	Dogwood Run #1 upstream water, whole	2.0	6.3	<.1	<.1	<.1	.7	2.0	<.1	--
D2AWS1	890706	Dogwood Run #2A septic downstream, filtered	--	--	.1	<.1	<.1	.4	3.0	<.1	--
D2AWS2	890706	Dogwood Run #2A septic downstream, whole	4.0	3.2	.1	<.1	<.1	.6	1.7	<.1	--
D2BWS1	890706	Dogwood Run #2B golf course trib, filtered	--	--	1.5	<.1	<.1	.6	12.3	<.1	--
D2BWS2	890706	Dogwood Run #2B golf course trib, whole	38	4.0	1.5	<.1	.1	.6	11.7	<.1	--
D3 WS1	890706	Dogwood Run #3 mixed downstream, filtered	--	--	.5	<.1	<.1	1.0	5.7	<.1	--
D3 WS2	890706	Dogwood Run #3 mixed downstream, whole	14	5.4	.5	<.1	<.1	1.2	5.7	<.1	--
D4 WS1	890706	Dogwood Run #4 sewage downstream, filtered	--	--	1.1	<.1	.1	.7	6.0	<.1	--
D4 WS2	890706	Dogwood Run #4 sewage downstream, whole	17	5.5	1.1	<.1	<.1	.7	7.0	<.1	--

Table B3. Concentrations of carbon (C), nitrogen (N), and sulfur (S) in N-source and fluvial samples from selected small watersheds in the Lower Susquehanna River Basin, Pennsylvania—Continued

[Concentrations are in parts per million (aqueous solutes are in milligrams per liter; solid compounds are in milligrams per kilogram); --, no data; <, less than; Replicate analysis results indicated by slash between values (for example, 1,120/1,180)]

Sample ID	Date sampled	Sample location and description	Carbon as C		Nitrogen as N				Sulfur as S		
			C-inorg	C-org	NO_3	NO_2	NH_3	N-org	Sulfate	Sulfide	S-org
		Samples collected during June 9 through July 6, 1989—Continued									
BE1WR1	890609	Bald Eagle runoff water, filtered	--	--	1.8	<0.1	<0.1	1.5	1.7	<0.1	--
BE1WR2	890609	Bald Eagle runoff water, whole	4.5	18	1.8	<.1	1.0	10.5	1.7	<.1	--
M1 WS1	890706	Monroe Creek #1 upstream water, filtered	--	--	.3	<.1	<.1	.4	1.3	<.1	--
M1 WS2	890706	Monroe Creek #1 upstream water, whole	4.0	5.3	.3	<.1	<.1	.6	1.3	<.1	--
M2AWS1	890706	Monroe Creek #2A midstream water, filtered	--	--	.4	<.1	<.1	.5	1.0	<.1	--
M2AWS2	890706	Monroe Creek #2A midstream water, whole	4.0	5.2	.7	<.1	<.1	.7	1.3	<.1	--
M2BWS1	890706	Monroe Creek #2B tributary water, filtered	--	--	1.5	<.1	14.0	.5	2.0	<.1	--
M2BWS2	890706	Monroe Creek #2B tributary water, whole	5.0	5.2	1.0	<.1	12.0	.5	1.3	<.1	--
M3 WS1	890706	Monroe Creek #3 downstream water, filtered	--	--	.4	<.1	<.1	.6	1.3	<.1	--
M3 WS2	890706	Monroe Creek #3 downstream water, whole	5.0	5.3	.5	<.1	<.1	.5	1.7	<.1	--
M2AWR1	890706	Monroe Creek 2A (runoff), filtered	--	--	<.1	<.1	<.1	1.0	1.3	<.1	--
M2AWR2	890706	Monroe Creek 2A (runoff), whole	10	10	<.1	<.1	<.1	.8	1.3	<.1	--
M2CWG1	890706	Monroe Creek 2C (spring), filtered	--	--	<.1	<.1	.1	.8	2.7	<.1	--
M2CWG2	890706	Monroe Creek 2C (spring), whole	38	13	1.0	<.1	.1	1.2	2.0	<.1	--
C1 WR1	890705	Conestoga FS#1 runoff water, filtered	--	--	1.8	<.1	.9	.2	3.7	<.1	--
C1 WR2	890705	Conestoga FS#1 runoff water, whole	43	3.1	.7	<.1	.5	.5	1.0	<.1	--
C1 WS1	890705	Conestoga FS#1 stream water, filtered	--	--	--	--	1.3	.6	4.7	<.1	--
C1 WS2	890705	Conestoga FS#1 stream water, whole	.8	6.9	--	--	1.2	1.2	4.3	<.1	--
DWAHW1	890706	Dillsburg Wastewater Plant effluent, filtered	--	--	5.2	<.1	<.1	.6	13.0	<.1	--
DWAHW2	890706	Dillsburg Wastewater Plant effluent, whole	39	3.0	5.2	<.1	<.1	.7	13.0	<.1	--
C1 WP1	890705	Conestoga FS#1 precipitation, filtered	--	--	.3	<.1	.6	.1	1.0	<.1	--
M2AWP1	890706	Monroe Creek precipitation, filtered	--	--	--	--	--	--	--	--	--

Table B3. Concentrations of carbon (C), nitrogen (N), and sulfur (S) in N-source and fluvial samples from selected small watersheds in the Lower Susquehanna River Basin, Pennsylvania—Continued

[Concentrations are in parts per million (aqueous solutes are in milligrams per liter; solid compounds are in milligrams per kilogram); --, no data; <, less than; Replicate analysis results indicated by slash between values (for example, 1,120/1,180)]

Sample ID	Date sampled	Sample location and description	Carbon as C		Nitrogen as N				Sulfur as S		
			C-inorg	C-org	NO_3	NO_2	NH_3	N-org	Sulfate	Sulfide	S-org
Samples collected during October 12, 1989, and January 26, 1990											
D1 S1	891012	Dogwood Run #1 topsoil	2,100	492,900	5,200	--	1,500	13,900	360	1,870	370
D1 WS1	891012	Dogwood Run #1 upstream water, filtered	--	--	<.1	--	<.1	.5	<1.0	<.1	--
D1 WS2	891012	Dogwood Run #1 upstream water, whole	2.1	3.5	<.1	--	<.1	.7	<1.0	<.1	--
D2AWS1	891012	Dogwood Run #2A septic downstream, filtered	--	--	.4	--	<.1	.5	2.0	<.1	--
D2AWS2	891012	Dogwood Run #2A septic downstream, whole	6.7	2.4	.4	--	<.1	.8	2.3	<.1	--
D2BWS1	891012	Dogwood Run #2B golf course tributary, filtered	--	--	3.1	--	<.1	.5	12.3	<.1	--
D2BWS2	891012	Dogwood Run #2B golf course tributary, whole	52	1.6	3.1	--	<.1	.7	12.0	<.1	--
D3 WS1	891012	Dogwood Run #3 mixed downstream, filtered	--	--	1.3	--	<.1	.6	6.0	<.1	--
D3 WS2	891012	Dogwood Run #3 mixed downstream, whole	28	1.9	1.3	--	<.1	.6	6.3	<.1	--
D4 WS1	891012	Dogwood Run #4 sewage downstream, filtered	--	--	2.0	--	<.1	1.6	9.7	<.1	--
D4 WS2	891012	Dogwood Run #4 sewage downstream, whole	31	3.0	2.0	--	<.1	2.0	10.0	<.1	--
DWAHW1	891012	Dillsburg Wastewater Plant effluent, filtered	--	--	4.8	--	<.1	.8	23.3	<.1	--
DWAHW2	891012	Dillsburg Wastewater Plant effluent, whole	38	4.3	5.0	--	<.1	1.5	22.7	<.1	--
BK1WG1	900126	Berks well water, upstream from septic, filtered	--	--	.1	--	<.1	.4	<1.0	<.1	--
BK1WG2	900126	Berks well water, upstream from septic, whole	9.8	<.5	.1	--	<.1	.4	<1.0	<.1	--
BK2HS1	900126	Berks septic field effluent, filtered	--	--	<.1	--	16.8	2.0	4.3	<.1	--
BK2HS2	900126	Berks septic field effluent, whole	29	14.	<.1	--	17.9	1.3	4.6	<.1	--

Table B4. Stable isotopes of carbon (C), nitrogen (N), and sulfur (S) in N-source and fluvial samples from selected small watersheds in the Lower Susquehanna River Basin, Pennsylvania

[Delta values are in per mil (‰) relative to standard reference materials ($\delta^{13}C$, Pee Dee Formation belemnite, PDB = 0.0 ‰; $\delta^{15}N$, atmospheric nitrogen, N_2 = 0.0 ‰; $\delta^{34}S$, Canyon Diablo troilite, CDT = 0.0 ‰; <, insufficient concentrations for analysis; --, not analyzed; $\delta^{15}N$ results reported for NO_3-N and then NH_3-N in parentheses, 7.7 (15.9); Replicate analysis results indicated by slash between values, 1.4/0.1; Sample was not homogeneous]

Sample ID	Date sampled	Sample location and description	$\delta^{13}C$	$\delta^{15}N$	$\delta^{34}S$
		Samples collected during June 27 through July 6, 1988			
BE1FL	880629	Bald Eagle synthetic fertilizer (liquid 30 percent N)	--	[1]–2.0 (–2.4)	5.8
BE1FS	880629	Bald Eagle synthetic fertilizer (solid 15-15-15)	–23.7	[2]1.4/.1	8.2
BE1S1	880629	Bald Eagle topsoil	–19.0	2.3	6.8/6.9
BE1S2	880629	Bald Eagle subsoil	–19.5	4.5	5.1
BR1MC	880706	Brush Run chicken manure	–17.8	1.1	[3]2.8/4.2
BR1MS1	880706	Brush Run swine manure filtered	--	11.4/10.6	5.8
BR1MS3	880706	Brush Run swine manure particulate	--	--	3.8
BR1S1	880706	Brush Run topsoil	–20.7	11.0	3.5
BR1S2	880706	Brush Run subsoil	–21.1	8.2	3.3
BR1WS1	880706	Brush Run stream water, filtered	--	5.2 (31.1)	11.2
BR1WS3	880706	Brush Run stream water, particulate	–24.1	7.6	7.7
CCPWS1	880630	Codorus Creek Pleasureville stream water, filtered	--	7.7 (15.9)	5.2
CCPWS3	880630	Codorus Creek Pleasureville stream water, particulate	–24.8	13.0	5.1
CCYWS1	880630	Codorus Creek York stream water, filtered	--	7.4/7.6 (5.3)	4.5
CCYWS3	880630	Codorus Creek York stream water, particulate	–24.5/–24.2	3.3/8.6	4.7
C1 MD	880628	Conestoga FS#1 dairy manure	–18.9	8.1	5.0
C1 S1	880628	Conestoga FS#1 topsoil	–22.1	4.0	–5.4
C1 S2	880628	Conestoga FS#1 subsoil	–20.4	5.2	2.6
C1 WS1	880628	Conestoga FS#1 stream water, filtered	--	7.2/7.5 (4.7)	5.1
C1 WS3	880628	Conestoga FS#1 stream water, particulate	–22.7	6.0	4.2

Table B4. Stable isotopes of carbon (C), nitrogen (N), and sulfur (S) in N-source and fluvial samples from selected small watersheds in the Lower Susquehanna River Basin, Pennsylvania—Continued

[Delta values are in per mil (‰) relative to standard reference materials ($\delta^{13}C$, Pee Dee Formation belemnite, PDB = 0.0 ‰; $\delta^{15}N$, atmospheric nitrogen, N_2 = 0.0 ‰; $\delta^{34}S$, Canyon Diablo troilite, CDT = 0.0 ‰; <, insufficient concentrations for analysis; --, not analyzed; $\delta^{15}N$ results reported for NO_3-N and then NH_3-N in parentheses, 7.7 (15.9); Replicate analysis results indicated by slash between values, 1.4/0.1; Sample was not homogeneous]

Sample ID	Date sampled	Sample location and description	$\delta^{13}C$	$\delta^{15}N$	$\delta^{34}S$
		Samples collected June 27 through July 6, 1988—Continued			
C2 MB	880628	Conestoga FS#2 steer manure	−12.2	9.5	5.8
C2 MS	880628	Conestoga FS#2 swine manure	−18.2	22.0	4.1
C2 MC	880628	Conestoga FS#2 chicken manure	−18.9/−18.6	3.2/3.1	3.8
C2 S1	880628	Conestoga FS#2 topsoil	−16.1	10.9	4.1
C2 S2	880628	Conestoga FS#2 subsoil	−18.0	9.8	3.5
C2 WG1	880628	Conestoga FS#2 spring water, filtered	--	8.2 (1.7)	3.7
C2 WG3	880628	Conestoga FS#2 stream water, particulate	−21.7	5.2	5.3
M1 FS	880706	Monroe Creek synthetic fertilizer (solid 34-6-7)	−20.7	4.3	−.6
M2AS1	880706	Monroe Creek topsoil	−24.4/−25.3	1.6/1.3	4.0
M2AS2	880706	Monroe Creek subsoil	−24.5/−24.6	3.5/3.4	4.8
M1 WS1	880706	Monroe Creek #1 upstream water, filtered	--	2.6 (4.4)	2.7
M1 WS3	880706	Monroe Creek #1 upstream water, particulate	−25.8	4.6	4.9
M2AWS1	880706	Monroe Creek #2 midstream water, filtered	--	7.9 (3.0)	4.6/5.1
M2AWS3	880706	Monroe Creek #2A midstream water, particulate	−24.3	6.9	4.5
SC S1	880627	Stony Creek topsoil	−27.8	.6	2.0
SC S2	880627	Stony Creek subsoil	−26.9	4.6	4.0
SC WS1	880627	Stony Creek stream water, filtered	--	3.8 (1.3)	4.4
SC WS3	880627	Stony Creek stream water, particulate	−27.3	.7	7.8
HWOHW1	880630	Harrisburg Wastewater Plant effluent filtered	--	1.6 (22.9/22.9)	5.7
HWOHW3	880630	Harrisburg Wastewater Plant effluent particulate	−23.4	−.9	6.1
HWOHW6	880630	Harrisburg Wastewater Plant primary sludge	−21.9	5.6	1.8/2.0
HWOHW7	880630	Harrisburg Wastewater Plant filter-press sludge	−21.7/−21.6	1.0/1.0	2.2
HW HS6	880630	Harrisburg Wastewater Plant septic-tank sludge	−24.1	−2.1	1.0
YWOHW1	880630	York Wastewater Plant O_2 effluent filtered	--	23.9 (18.0)	6.4
YWOHW3	880630	York Wastewater Plant O_2 effluent particulate	−24.3	2.8	4.7
YWDHW1	880630	York Wastewater Plant direct effluent filtered	--	9.8 (18.4)	5.9
YWDHW3	880630	York Wastewater Plant direct effluent particulate	−20.8	1.6	4.7
YW HW6	880630	York Wastewater Plant primary sludge	−21.4	10.1	2.4

Table B4. Stable isotopes of carbon (C), nitrogen (N), and sulfur (S) in N-source and fluvial samples from selected small watersheds in the Lower Susquehanna River Basin, Pennsylvania—Continued

[Delta values are in per mil (‰) relative to standard reference materials ($\delta^{13}C$, Pee Dee Formation belemnite, PDB = 0.0 ‰; $\delta^{15}N$, atmospheric nitrogen, N_2 = 0.0 ‰; $\delta^{34}S$, Canyon Diablo troilite, CDT = 0.0 ‰; <, insufficient concentrations for analysis; --, not analyzed; $\delta^{15}N$ results reported for NO_3-N and then NH_3-N in parentheses, 7.7 (15.9); Replicate analysis results indicated by slash between values, 1.4/0.1; Sample was not homogeneous]

Sample ID	Date sampled	Sample location and description	$\delta^{13}C$	$\delta^{15}N$	$\delta^{34}S$
		Samples collected December 14, 1988, through January 5, 1989			
BE1S1	881214	Bald Eagle topsoil	−19.5	3.6	6.0
BE1S2	881214	Bald Eagle subsoil	−20.1/−20.0	4.3/4.9	6.5
BE2WS1	881214	Bald Eagle downstream water, filtered	--	5.8	9.9
BE2WS3	881214	Bald Eagle downstream water, particulate	−25.4	5.0	9.9
BH1WS1	881222	Berkshire Hills #1 septic upstream, filtered	--	4.1	4.4
BH1WS3	881222	Berkshire Hills #1 stream water, particulate	−31.4/−30.9	3.0/2.3	--
BH2WS1	881222	Berkshire Hills #2 septic downstream, filtered	--	4.2	4.7
BH2WS3	881222	Berkshire Hills #2 stream water, particulate	−27.1	3.4	4.4
BR1MS4	881214	Brush Run swine feces	−15.0/−15.0	6.2/6.4	3.6
BR1MS1	881214	Brush Run swine manure filtered	--	12.4 (37.2)	4.9
BR1MS3	881214	Brush Run swine manure particulate	−14.9	7.7	3.8/3.6
BR1S1	881214	Brush Run topsoil	−19.1	12.4/12.3	3.6
BR1S2	881214	Brush Run subsoil	−19.7	11.2	4.1
BR1WS1	881214	Brush Run stream water, filtered	--	39.3 (17.2)	4.8/4.8
BR1WS3	881214	Brush Run stream water, particulate	−20.3	5.0	--
D1 S1	881215	Dogwood Run #1 topsoil	−27.4	.0	3.0
D1 S2	881215	Dogwood Run #1 subsoil	−26.5	5.3	9.8
D1 WS1	881215	Dogwood Run #1 upstream water, filtered	--	2.2	3.6
D1 WS3	881215	Dogwood Run #1 upstream water, particulate	−27.0	1.7	5.1
D2AWS1	881215	Dogwood Run #2A septic downstream, filtered	--	3.9	4.5
D2AWS3	881215	Dogwood Run #2A septic downstream, particulate	−26.7	3.0	5.5
D3 WS1	881215	Dogwood Run #3 mixed downstream, filtered	--	6.9	8.7
D3 WS3	881215	Dogwood Run #3 mixed downstream, particulate	−29.3	3.8	4.3
D4 WS1	881215	Dogwood Run #4 sewage downstream, filtered	--	15.7 (4.4)	5.3
D4 WS3	881215	Dogwood Run #4 sewage downstream, particulate	−23.4	9.2	4.6

Table B4. Stable isotopes of carbon (C), nitrogen (N), and sulfur (S) in N-source and fluvial samples from selected small watersheds in the Lower Susquehanna River Basin, Pennsylvania—Continued

[Delta values are in per mil (‰) relative to standard reference materials ($\delta^{13}C$, Pee Dee Formation belemnite, PDB = 0.0 ‰; $\delta^{15}N$, atmospheric nitrogen, N_2 = 0.0 ‰; $\delta^{34}S$, Canyon Diablo troilite, CDT = 0.0 ‰; <, insufficient concentrations for analysis; --, not analyzed; $\delta^{15}N$ results reported for NO_3-N and then NH_3-N in parentheses, 7.7 (15.9); Replicate analysis results indicated by slash between values, 1.4/0.1; Sample was not homogeneous]

Sample ID	Date sampled	Sample location and description	$\delta^{13}C$	$\delta^{15}N$	$\delta^{34}S$
		Samples collected December 14, 1988, through January 5, 1989—Continued			
CLMMD	881221	Conestoga LM dairy manure	−20.1	9.1	4.4
C1 MD	881221	Conestoga FS#1 dairy manure	−18.8/−17.9	8.7/8.6	3.4
C1 S1	881221	Conestoga FS#1 topsoil	−18.4	6.9	4.8
C1 S2	881221	Conestoga FS#1 subsoil	−18.0	7.0	6.6
C1 WS1	881221	Conestoga FS#1 stream water, filtered	--	5.0	5.0
C1 WS3	881221	Conestoga FS#1 stream water, particulate	−25.6	5.0	4.7
C2 MB	881221	Conestoga FS#2 steer manure	−14.2	11.6	5.4
C2 MC	881221	Conestoga FS#2 chicken manure	−17.9/−18.0	1.5/1.7	2.3
C2 MS1	881221	Conestoga FS#2 swine manure filtered	--	-- (--)	--
C2 MS3	881221	Conestoga FS#2 swine manure particulate	−17.0/−17.0	9.5/9.9	4.3
C2 S1	881221	Conestoga FS#2 topsoil	−18.5	10.8	6.7/6.4
C2 S2	881221	Conestoga FS#2 subsoil	−19.0	9.1	6.6/8.6
C2 WR1	881221	Conestoga FS#2 runoff water, filtered	--	-- (12.9)	2.7
C2 WR3	881221	Conestoga FS#2 runoff water, particulate	−20.4	6.4	4.4
SC S1	881222	Stony Creek topsoil	−27.3	.3	3.5
SC S2	881222	Stony Creek subsoil	−26.9	4.6	5.8
SC WS1	881222	Stony Creek stream water, filtered	--	4.3	4.4
SC WS3	881222	Stony Creek stream water, particulate	−27.7	1.2	4.3
DWAHW1	881215	Dillsburg Wastewater Plant effluent filtered	--	26.7 (2.4)	3.5
DWAHW3	881215	Dillsburg Wastewater Plant effluent particulate	−22.3	10.2	5.1
DWAHW6	881215	Dillsburg Wastewater Plant sludge particulate	−22.6/−22.6	11.1/11.1	3.4
HW HS1	890105	Harrisburg Wastewater Plant septic tank filtered	--	3.3 (31.2)	9.0
HW HS3	890105	Harrisburg Wastewater Plant septic susp. sludge	−21.6/−22.2	1.0/1.3	4.3/4.0

Table B4. Stable isotopes of carbon (C), nitrogen (N), and sulfur (S) in N-source and fluvial samples from selected small watersheds in the Lower Susquehanna River Basin, Pennsylvania—Continued

[Delta values are in per mil (‰) relative to standard reference materials ($\delta^{13}C$, Pee Dee Formation belemnite, PDB = 0.0 ‰; $\delta^{15}N$, atmospheric nitrogen, N_2 = 0.0 ‰; $\delta^{34}S$, Canyon Diablo troilite, CDT = 0.0 ‰; <, insufficient concentrations for analysis; --, not analyzed; $\delta^{15}N$ results reported for NO_3-N and then NH_3-N in parentheses, 7.7 (15.9); Replicate analysis results indicated by slash between values, 1.4/0.1; Sample was not homogeneous]

Sample ID	Date sampled	Sample location and description	$\delta^{13}C$	$\delta^{15}N$	$\delta^{34}S$
		Samples collected April 27 through May 24, 1989			
BE1FS	890505	Bald Eagle synthetic fertilizer (solid 15-15-15)	−22.9/−24.1	−0.5/−.4	8.3
BE1S1	890505	Bald Eagle topsoil	−20.7	2.5	3.3
BE1S2	890505	Bald Eagle subsoil	−19.7	3.5	6.6
BE1WR1	890505	Bald Eagle runoff water, filtered	--	−.5 (8.6)	3.6/3.6
BE1WR3	890505	Bald Eagle runoff water, particulate	−18.6	5.4	4.5/5.2
BE2WS1	890505	Bald Eagle stream water, filtered	--	6.2/6.5	5.3
BE2WS3	890505	Bald Eagle stream water, particulate	−24.2	7.4	4.8/4.5
BE2WS1	890524	Bald Eagle stream water, filtered	--	6.2	5.7
BE2WS3	890524	Bald Eagle stream water, particulate	−22.3	6.1	5.3/5.0
BH1WS1	890518	Berkshire Hills #1 septic upstream, filtered	--	4.4	6.0
BH1WS3	890518	Berkshire Hills #1 septic upstream, particulate	−20.7	8.6	3.1/3.3
BH2WS1	890518	Berkshire Hills #2 septic downstream, filtered	--	4.9	4.7
BH2WS3	890518	Berkshire Hills #2 septic downstream, particulate	−27.2	3.7	--
BR1MS1	890505	Brush Run swine manure filtered	--	< (33.8)	−.9
BR2MS3	890505	Brush Run swine manure particulate	−15.0	12.5	4.5/4.3
BR1S1	890505	Brush Run topsoil	−19.1	8.7	4.5
BR1S2	890505	Brush Run subsoil	−19.0	7.7	4.1
BR2WS1	890505	Brush Run stream water (pond), filtered	--	16.2 (17.1)	6.7
BR2WS3	890505	Brush Run stream water (pond), particulate	−20.8	10.2	5.2

Table B4. Stable isotopes of carbon (C), nitrogen (N), and sulfur (S) in N-source and fluvial samples from selected small watersheds in the Lower Susquehanna River Basin, Pennsylvania—Continued

[Delta values are in per mil (‰) relative to standard reference materials ($\delta^{13}C$, Pee Dee Formation belemnite, PDB = 0.0 ‰; $\delta^{15}N$, atmospheric nitrogen, N_2 = 0.0 ‰; $\delta^{34}S$, Canyon Diablo troilite, CDT = 0.0 ‰; <, insufficient concentrations for analysis; --, not analyzed; $\delta^{15}N$ results reported for NO_3-N and then NH_3-N in parentheses, 7.7 (15.9); Replicate analysis results indicated by slash between values, 1.4/0.1; Sample was not homogeneous]

Sample ID	Date sampled	Sample location and description	$\delta^{13}C$	$\delta^{15}N$	$\delta^{34}S$
		Samples collected April 27 through May 24, 1989—Continued			
D1 S1	890503	Dogwood Run #1 topsoil	−26.7/−26.7	0.5/.7	4.4
D1 S2	890503	Dogwood Run #1 subsoil	−26.2	5.1	6.0
D1 WS1	890518	Dogwood Run #1 upstream water, filtered	--	<	4.9
D1 WS3	890518	Dogwood Run #1 upstream water, particulate	−26.9	−1.1	2.6
D2AWS1	890518	Dogwood Run #2A septic downstream, filtered	--	<	--
D2AWS3	890518	Dogwood Run #2A septic downstream, particulate	−27.5	.4	6.1
D2BWS1	890518	Dogwood Run #2B golf course trib, filtered	--	5.4	12.7
D2BWS3	890518	Dogwood Run #2B golf course trib, particulate	−27.6	2.3	13.0 •
D3 WS1	890518	Dogwood Run #3 mixed downstream, filtered	--	5.2	8.7/8.9
D3 WS3	890518	Dogwood Run #3 mixed downstream, particulate	−27.5	.7	2.5
D4 WS1	890518	Dogwood Run #4 sewage downstream, filtered	--	5.9/6.1	6.3
D4 WS3	890518	Dogwood Run #4 sewage downstream, particulate	−25.6	5.5	3.1
SC S1	890503	Stony Creek topsoil	−26.4	.4	4.0/3.8
SC S2	890503	Stony Creek subsoil	−26.1	3.9	6.4
SC WS1	890503	Stony Creek stream water, filtered	--	<	4.0
SC WS3	890503	Stony Creek stream water, particulate	−27.7	.9	3.1
CCPWS1	890504	Codorus Creek Pleasureville stream water, filtered	--	9.3	.8
CCPWS3	890504	Codorus Creek Pleasureville stream water, particulate	−24.8/−25.1	5.3/5.6	2.6/2.5
CCYWS1	890504	Codorus Creek York stream water, filtered	--	6.8	4.2
CCYWS3	890504	Codorus Creek York stream water, particulate	−25.7	4.9	2.9/3.3
C1 MD	890427	Conestoga FS#1 dairy manure	−20.6	3.8	3.7/4.3/4.1
C1 S1	890427	Conestoga FS#1 topsoil	−20.7	5.3	5.5
C1 S2	890427	Conestoga FS#1 subsoil	−21.2	1.8	6.1/6.1/5.6
C1 WR1	890502	Conestoga FS#1 runoff water, filtered	--	8.5	--
C1 WR3	890502	Conestoga FS#1 runoff water, particulate	--	7.8/7.8	2.4
C1 WS1	890502	Conestoga FS#1 stream water (high flow), filtered	--	--	4.6
C1 WS3	890502	Conestoga FS#1 stream water (high flow), particulate	−23.9	4.7	3.9/3.4
C1 WS1	890427	Conestoga FS#1 stream water (low flow), filtered	--	--	4.9
C1 WS3	890427	Conestoga FS#1 stream water (low flow), particulate	−23.4	4.7	5.3

Table B4. Stable isotopes of carbon (C), nitrogen (N), and sulfur (S) in N-source and fluvial samples from selected small watersheds in the Lower Susquehanna River Basin, Pennsylvania—Continued

[Delta values are in per mil (‰) relative to standard reference materials ($\delta^{13}C$, Pee Dee Formation belemnite, PDB = 0.0 ‰; $\delta^{15}N$, atmospheric nitrogen, N_2 = 0.0 ‰; $\delta^{34}S$, Canyon Diablo troilite, CDT = 0.0 ‰; <, insufficient concentrations for analysis; --, not analyzed; $\delta^{15}N$ results reported for NO_3-N and then NH_3-N in parentheses, 7.7 (15.9); Replicate analysis results indicated by slash between values, 1.4/0.1; Sample was not homogeneous]

Sample ID	Date sampled	Sample location and description	$\delta^{13}C$	$\delta^{15}N$	$\delta^{34}S$
		Samples collected April 27 through May 24, 1989—Continued			
C2 MB	890428	Conestoga FS#2 steer manure	−14.0	11.3	6.8
C2 MC	890428	Conestoga FS#2 chicken manure	−18.3	10.2	3.8
C2 MS1	890428	Conestoga FS#2 swine manure filtered	--	<	--
C2 MS3	890428	Conestoga FS#2 swine manure particulate	−18.4	14.8	3.2
C2 S1	890428	Conestoga FS#2 topsoil	−18.2	7.4	6.3
C2 S2	890428	Conestoga FS#2 subsoil	−21.4	5.6	6.9
C2 WG1	890428	Conestoga FS#2 spring water, filtered	--	10.7 (−2.5)	3.7
C2 WG3	890428	Conestoga FS#2 spring water, particulate	−26.6	6.1	4.0
C2AWR1	890502	Conestoga FS#2 runoff #1 water, filtered	--	4.2 (3.5)	.3
C2AWR3	890502	Conestoga FS#2 runoff #1 water, particulate	−26.1/−26.1	7.8/7.7	3.8/3.5
C2BWR1	890502	Conestoga FS#2 runoff #2 water, filtered	--	.3 (42.1)	4.0
C2BWR3	890502	Conestoga FS#2 runoff #2 water, particulate	−17.9	11.8	4.1/3.7
M2AFS	880706	Monroe Creek synthetic fertilizer (solid 34-6-7)	−20.5	1.9	−.4
M2AFS1	890510	Monroe Creek synthetic fertilizer (solid 34-3-8)	−42.5	−.9	19.9
M2AFS2	890510	Monroe Creek synthetic fertilizer (solid 18-4-10)	−41.3	−.2	7.6
M2AS1	890502	Monroe Creek topsoil	−24.2	4.0	4.0
M2AS2	890502	Monroe Creek subsoil	−24.0/−24.1	4.5/4.8	5.3
M1 WS1	890510	Monroe Creek #1 upstream water, filtered	--	8.7	4.3
M1 WS3	890510	Monroe Creek #1 upstream water, particulate	−24.8	4.1	--
M2AWS1	890510	Monroe Creek #2A midstream water, filtered	--	9.2	4.2
M2AWS3	890510	Monroe Creek #2A midstream water, particulate	−25.6	4.2	--
M3 WS1	890510	Monroe Creek #3 downstream water, filtered	--	6.9	4.4
M3 WS3	890510	Monroe Creek #3 downstream water, particulate	−25.4/−25.0	3.2/4.0	3.3
M2AWR1	890510	Monroe Creek #2A runoff water, filtered	--	<	--
M2AWR3	890510	Monroe Creek #2A runoff water, particulate	−28.1/−27.4	.9/1.3	8.2
M2CWG1	890510	Monroe Creek #2C spring water, filtered	--	<	4.8
M2CWG3	890510	Monroe Creek #2C spring water, particulate	−26.1/−23.8	5.9/1.1	--

Table B4. Stable isotopes of carbon (C), nitrogen (N), and sulfur (S) in N-source and fluvial samples from selected small watersheds in the Lower Susquehanna River Basin, Pennsylvania—Continued

[Delta values are in per mil (‰) relative to standard reference materials ($\delta^{13}C$, Pee Dee Formation belemnite, PDB = 0.0 ‰; $\delta^{15}N$, atmospheric nitrogen, N_2 = 0.0 ‰; $\delta^{34}S$, Canyon Diablo troilite, CDT = 0.0 ‰; <, insufficient concentrations for analysis; --, not analyzed; $\delta^{15}N$ results reported for NO_3-N and then NH_3-N in parentheses, 7.7 (15.9); Replicate analysis results indicated by slash between values, 1.4/0.1; Sample was not homogeneous]

Sample ID	Date sampled	Sample location and description	$\delta^{13}C$	$\delta^{15}N$	$\delta^{34}S$
		Samples collected April 27 through May 24, 1989—Continued			
DWAHW1	890518	Dillsburg Wastewater Plant effluent filtered	--	--	4.5
DWAHW3	890518	Dillsburg Wastewater Plant activated sludge	−22.3	7.9	4.8
HW HS1	890505	Harrisburg Wastewater Plant septic tank filtered	--	−.7 (1.4)	--
HW HS3	890505	Harrisburg Wastewater Plant septic tank sludge	−22.7	3.0	3.4/4.7/4.1
HWOHW1	890504	Harrisburg Wastewater Plant effluent filtered	--	11.7 (19.3)	5.7
HWOHW3	890504	Harrisburg Wastewater Plant effluent particulate	−21.4	−3.7	3.7
YWRHW1	890504	York Wastewater Plant effluent filtered	--	13.7	3.5
YWRHW3	890504	York Wastewater Plant effluent particulate	−21.4	8.3	5.4
BE2WP1	890502	Bald Eagle Precipitation, filtered—boiled	--	<	<
BR1WP1	890502	Brush Run Precipitation, filtered—boiled	--	--	−1.1
C1 WP1	890505	Conestoga FS#1 Precipitation, filtered—boiled	--	<	<
C2 WP1	890505	Conestoga FS#2 Precipitation, filtered—boiled	--	<	<
M2AWP1	890502	Monroe Creek Precipitation, filtered—boiled	--	<	<
M2AWP1	890510	Monroe Creek Precipitation, filtered	--	<	<
		Samples collected June 9 through July 6, 1989			
D1 WS1	890706	Dogwood Run #1 upstream water, filtered	--	<	4.6
D1 WS3	890706	Dogwood Run #1 upstream water, particulate	−27.3/−26.4	.6/−.2	4.5
D2AWS1	890706	Dogwood Run #2A septic downstream water, filtered	--	4.2	6.1
D2AWS3	890706	Dogwood Run #2A septic downstream water, particulate	−27.0/−26.8	1.5/2.2	4.5
D2BWS1	890706	Dogwood Run #2B golf course trib. water, filtered	--	6.0/5.7	12.6
D2BWS3	890706	Dogwood Run #2B golf course trib. water, particulate	−27.1/−27.1	3.6/4.1	5.6
D3 WS1	890706	Dogwood Run #3 mixed downstream water, filtered	--	4.9	10.3
D3 WS3	890706	Dogwood Run #3 mixed downstream water, particulate	−26.9/−26.8	2.7/2.5	4.8
D4 WS1	890706	Dogwood Run #4 sewage downstream water, filtered	--	--	8.0
D4 WS3	890706	Dogwood Run #4 sewage downstream water, particulate	−25.7	5.0	2.5

Table B4. Stable isotopes of carbon (C), nitrogen (N), and sulfur (S) in N-source and fluvial samples from selected small watersheds in the Lower Susquehanna River Basin, Pennsylvania—Continued

[Delta values are in per mil (‰) relative to standard reference materials ($\delta^{13}C$, Pee Dee Formation belemnite, PDB = 0.0 ‰; $\delta^{15}N$, atmospheric nitrogen, N_2 = 0.0 ‰; $\delta^{34}S$, Canyon Diablo troilite, CDT = 0.0 ‰; <, insufficient concentrations for analysis; --, not analyzed; $\delta^{15}N$ results reported for NO_3-N and then NH_3-N in parentheses, 7.7 (15.9); Replicate analysis results indicated by slash between values, 1.4/0.1; Sample was not homogeneous]

Sample ID	Date sampled	Sample location and description	$\delta^{13}C$	$\delta^{15}N$	$\delta^{34}S$
		Samples collected during June 9 through July 6, 1989—Continued			
BE1WR1	890609	Bald Eagle runoff water, filtered	--	4.7	3.6
BE1WR3	890609	Bald Eagle runoff water, particulate	−27.1	3.8	5.5
M1 WS1	890706	Monroe Creek #1 upstream water, filtered	--	<	4.7
M1 WS3	890706	Monroe Creek #1 upstream water, particulate	−26.3	1.1	3.6
M2AWS1	890706	Monroe Creek #2A midstream water, filtered	--	5.8	3.6
M2AWS3	890706	Monroe Creek #2A midstream water, particulate	−26.1	2.2	3.7
M2BWS1	890706	Monroe Creek #2B tributary water, filtered	--	5.8 (7.3)	5.4
M2BWS3	890706	Monroe Creek #2B tributary water, particulate	−25.5	5.2	--
M3 WS1	890706	Monroe Creek #3 downstream water, filtered	--	6.6	4.9
M3 WS3	890706	Monroe Creek #3 downstream water, particulate	−25.7	2.0	5.1
M2AWR1	890706	Monroe Creek #2A runoff water, filtered	--	<	8.4
M2AWR3	890706	Monroe Creek #2A runoff water, particulate	−25.2	5.2	4.8
M2CWG1	890706	Monroe Creek #2C spring water, filtered	--	<	6.3
M2CWG3	890706	Monroe Creek #2C spring water, particulate	−29.1/−29.0	4.9/4.9	4.8
C1 WR1	890705	Conestoga FS#1 runoff water, filtered	--	−.2 (.2)	--
C1 WR3	890705	Conestoga FS#1 runoff water, particulate	−19.8	8.3	3.3/4.2
C1 WS1	890705	Conestoga FS#1 stream water, filtered	--	<	3.2
C1 WS3	890705	Conestoga FS#1 stream water, particulate	−21.7	5.4	4.2
DWAHW1	890706	Dillsburg Wastewater Plant effluent filtered	--	<	4.6
DWAHW3	890706	Dillsburg Wastewater Plant effluent particulate	−22.2	7.4	3.6
C1 WP1	890705	Conestoga FS#1 Precipitation, filtered	--	<	4.5
M2AWP1	890706	Monroe Creek Precipitation, filtered	--	<	<

Table B4. Stable isotopes of carbon (C), nitrogen (N), and sulfur (S) in N-source and fluvial samples from selected small watersheds in the Lower Susquehanna River Basin, Pennsylvania—Continued

[Delta values are in per mil (‰) relative to standard reference materials ($\delta^{13}C$, Pee Dee Formation belemnite, PDB = 0.0 ‰; $\delta^{15}N$, atmospheric nitrogen, N_2 = 0.0 ‰; $\delta^{34}S$, Canyon Diablo troilite, CDT = 0.0 ‰; <, insufficient concentrations for analysis; --, not analyzed; $\delta^{15}N$ results reported for NO_3-N and then NH_3-N in parentheses, 7.7 (15.9); Replicate analysis results indicated by slash between values, 1.4/0.1; Sample was not homogeneous]

Sample ID	Date sampled	Sample location and description	$\delta^{13}C$	$\delta^{15}N$	$\delta^{34}S$
		Samples collected October 12, 1989, and January 26, 1990			
D1 S1	891012	Dogwood Run #1 topsoil	−27.3	−.3	4.7/4.6
D1 WS1	891012	Dogwood Run #1 upstream water, filtered	--	<	1.7
D1 WS3	891012	Dogwood Run #1 upstream water, particulate	−27.7	1.3	4.8
D2AWS1	891012	Dogwood Run #2A septic downstream water, filtered	--	<	7.0
D2AWS3	891012	Dogwood Run #2A septic downstream water, particulate	−26.6/−26.3	2.9	2.3
D2BWS1	891012	Dogwood Run #2B golf course trib. water, filtered	--	<	10.3
D2BWS3	891012	Dogwood Run #2B golf course trib. water, particulate	−27.9	4.5	5.5
D3 WS1	891012	Dogwood Run #3 mixed downstream water, filtered	--	<	9.6
D3 WS3	891012	Dogwood Run #3 mixed downstream water, particulate	−27.4	4.5	4.6
D4 WS1	891012	Dogwood Run #4 sewage downstream water, filtered	--	<	6.9
D4 WS3	891012	Dogwood Run #4 sewage downstream water, particulate	−23.2/−23.2	6.7/6.9	4.7
DWAHW1	891012	Dillsburg Wastewater Plant effluent filtered	--	<	--
DWAHW3	891012	Dillsburg Wastewater Plant effluent particulate	−21.6	8.4/7.7	4.9
BK1WG1	900126	Berks well water, filtered	--	<	<
BK1WG3	900126	Berks well water, particulate	−30.3	−1.6	<
BK2HS1	900126	Berks septic field effluent, filtered	--	<	9.4
BK2HS3	900126	Berks septic field effluent, particulate	−23.4/−23.4	−.6/−.9	3.5

*U.S. Government Printing Office: 2002 — 746-721